Glossary of Structural Geology and Tectonics

Glossary of Structural Geology and Tectonics

Edited by :

P.S. Saklani

Department of Geology
University of Delhi (Retired)
and
Emeritus Professor
Netaji Subhas Institute of Technology
Dwarka, New Delhi - 110075

2008

SSPH

SATISH SERIAL PUBLISHING HOUSE

403, Express Tower, Commercial Complex, Azadpur, Delhi-110033 (India)
Phone : 011-27672469, 27672852, Fax : 91-11-27672046
E-mail :info@satishserial.com, hkjain1975@yahoo.com
Website : www.satishserial.com

Published by :

SATISH SERIAL PUBLISHING HOUSE
403, Express Tower, Commercial Complex, Azadpur,
Delhi-110033 (India)
Phone : 011-27672852, Fax : 91-11-27672046
E-mail : info@satishserial.com, hkjain1975@yahoo.com
Website : www.satishserial.com

First Published 2008

ISBN : 81-89304-44-5

Printed at

SALASAR IMAGING SYSTEMS, DELHI-35

PREFACE

I have been teaching Structural Geology and Tectonics for the past four decades and during this period, I always felt a need for a glossary or definitional dictionary devoted to this branch. I therefore, made an attempt in this direction.

The editor for while doing the groundwork of the book made use of Glossary of Geology edited by R. Bates & J. Jackson (Am. Geol. Inst., 1980); Glossary of Geology in Hindi edited by myself (CSTT Govt. of India 1996); Introduction to the structure of the earth by E. Spencer; Tectonic Geology by myself (2006); Geology: an Introduction by myself (2004).

This glossary is based on the works of many structural geologists and I have always tried to acknowledge their works. I have attempted to explain many terms with the help of illustrations citing many Indian examples wherever possible.

During the preparation of this book. I held discussions with my students namely, Dr. Satendra (New Delhi); Dr. D.C. Nainwal (Gopeshwar); Dr. S.C. Bhatt, Dr. B.C. Joshi, Dr. V.K. Singh and Prof. S.P. Singh (Bundelkhand University, Jhansi); Prof. A.K. Shandilya (Sagar University); Prof. C.S. Dubey, K.N. Kandwal (Delhi University); S. Shekhar (Ground Water Board, New Delhi); Dr. A.C. Pandey (BIT Mesra - Ranchi) and many figures have been taken in this book from their research works.

The book is being published under the aegis of the Geoscience Foundation, India and hopefully it would be useful to students, research scholars and teachers connected with this branch of geology.

P.S. Saklani
Secretary General
Geoscience Foundation, India.

CONTENTS

A

A Axis It is an othogonal reference axis, It describes the geometry of a fabric possessing monoclinic symmetry. In a monoclinic symmetry, it represents progressive simple shear and as such the *a axis (= x- axis)* again lies in the unique plane of symmetry but parallel to the movement plane. It is commonly referred to as the direction of tectonic transport. In a progressive simple shear the *a axis* is the direction of shear. Some workers refer the a, b and c axes as x,y and z. (Fig. 1)

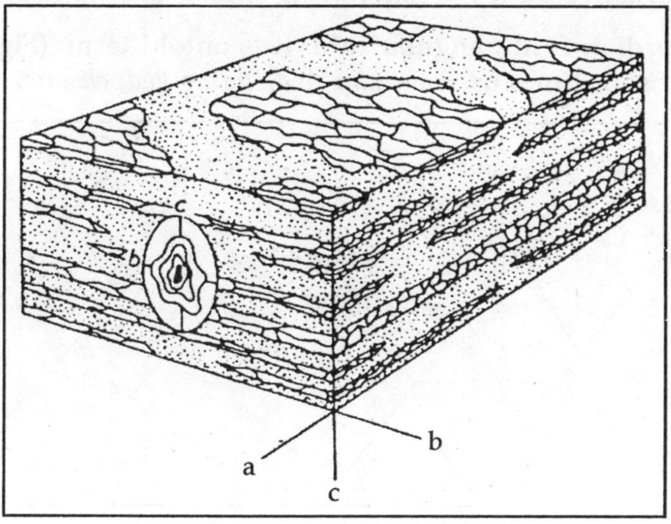

Fig. 1. Tectonite and its a, b and c axes

Anticlinorium It is anticlinorium in which the axial surfaces of the subsidiary folds converge upwards. While in a synclinorium the axial surfaces of the subsidiary folds converge downwards. (Fig. 2).

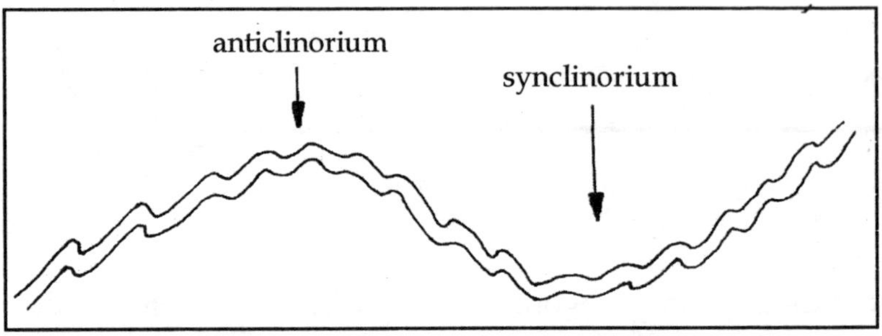

Fig. 2. Anticlinorium : Synclinorium

Acadian orogeny A Middle Palaeozoic deformation, especially in the northern Appalachians. It is named after Acadia, (old French name for the Canadian Maritime Provinces). It has been dated radiometrically as between 330 and 360 m.y. ago.

Accordant fold It is a fold having similar orientation.

Accordian fold : angular fold. It is an old term, (Fig. 3) sometimes used for *kink fold, zigzag fold; chevron fold.*

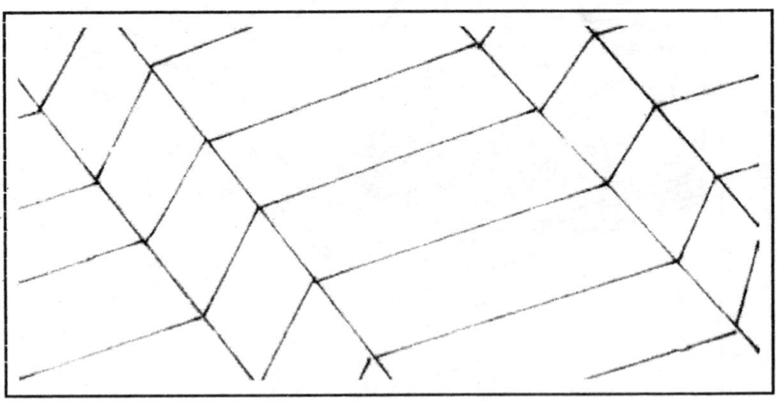

Fig. 3. Accordian Fold

Accreting plate boundary A boundary between the *two moving apart plates,* with formation of new oceanic-type lithosphere (Dennis & Atwater, 1974, Dennis (ed), 1967).

Accretion A sedimentary process by which an inorganic body increases in size by the external addition of fresh particles with deposition of eolian sand on a continuous sand surface.

Acrobatholithic A little used term. It is a mineral deposit occurring in or near an exposed batholith dome; also, due to batholith erosion (Emmons, 1933).

Acitve fault A fault which is characterized by recurrent movement due to the periodic displacements or seismic activity.

A direction = *a axis.*

Advection Lateral mass movements of mantle material. *Compare : convection.*

Aerial mapping Maps prepared by aerial photographs for interpretation of geology.

Affine A homegeneous deformation, i.e. straight lines remain straight after deformation (Fig.4).

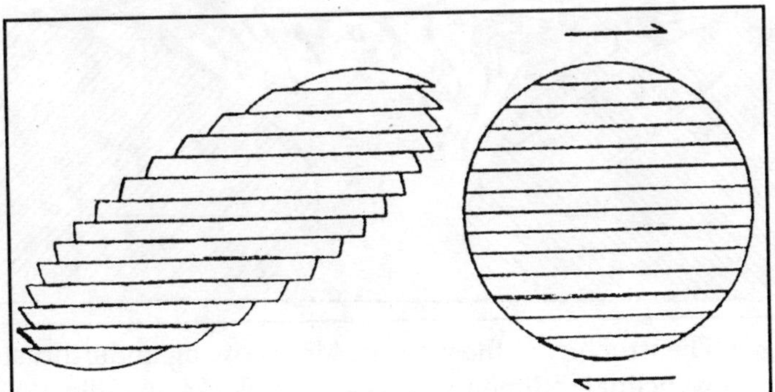

Fig. 4. Diagrammatic example of a circle changed into an ellipse by movement along glide planes. (After Billings, 2000).

Aftershock An earthquake ocurring after a large earthquake *(main shock)* near the focus of the larger earthquake. Aftershocks, decrease in frequency and magnitude with time.

Airy hypothesis This mechanism of *isostasy*, was proposed by George Bedell Airy, indicating an equilibrium of crustal blocks of the same density but of different size.

Akmolith : acmolith An igneous intrusion occurring along a zone of décollement, with or without extension.

Allegheny orogeny An orogeny characterized by deformed rocks of the Valley and Ridge province of the adjacent Allegheny Plateau of the centeral and southern Appalachians.

Allochthon : allochthone A mass of rock which is not in its place of origin due to tectonic movement, as is also observed in a *thurst sheet* or *nappe*.

Allochthone : *allochthon.*

Allochthonous : allogenic Transported elsewhere than its present place; of foreign origin, transported on a low-angle thrust fault. The term was first used by Naumann (1858) for rocks of distant origin. (Fig. 5).

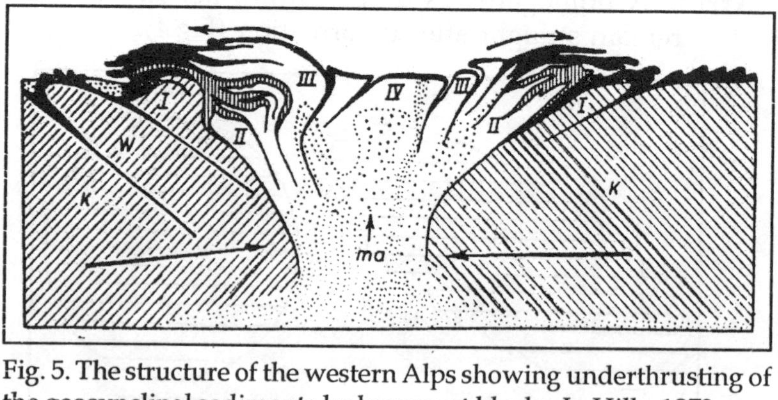

Fig. 5. The structure of the western Alps showing underthrusting of the geosynclinal sediments by basement blocks. In Hills, 1973.

K, crystalline basement; *W*, crystalline wedge; *I*, autochthonous nappes; *II*, parautochthonous nappes; *III*, exotic nappes; *IV*, median massif; ma, granitic magma. (*After Kober*)

Alpides : Alpine-Himalayan belt A name used by Suess for the great orogenic belt or system of young folded mountains, including the Alps, extending east ward from Spain into southern Asia.

Alpine The structural features resembling in complexity with those of the European Alps, regardless of the age or location of the mountains.

Alpine orogeny A young orogenic event of European and Asian region, by which the rocks of the Alps (Alpides) were strongly deformed. Many geologists restrict this into the Tertiary i.e., during the Miocene or Pliocene. (Fig. 6)

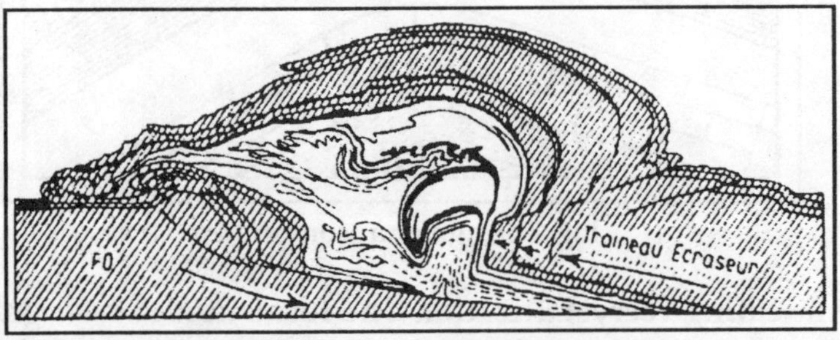

Fig. 6. The nappe structures of the Western Alps, showing the Traineau ecraseur pushing against the Alpine edifice and over-riding the foreland of Europe (Fo). (After Agrand, in Hills, 1973)

Alpinotype tectonics The tectonic features of orogenic belts. Their internal parts are characterized by deeps-seated folding and plutonism, The external parts show lateral thrusting, associated with the nappes / thrust sheets. (Fig. 6).

Amplitude [fold] A symmetrical fold system having an amplitude or a wave form, i.e. it has half the orthogonal distance between the antiformal and synformal *enveloping surfaces*.

Angular fold A kink fold which has a less angular hinge.

Angular unconformity : *angular discordance; clinonconformity; structural unconformity; orogenic unconformity.* An *unconformity* between the two groups of rocks. The older, underlying rocks dip at a diffferent angle(usually steeper) than the younger overlying strata resting upon the eroded surface of tilted or folded older rocks. (Fig. 7)

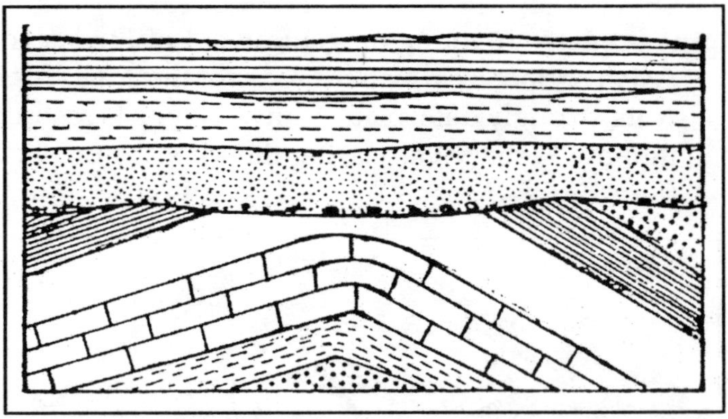

Fig. 7. Angular unconformity

Annealing recrystallization The formation of new grains in a rock formed in soild-state deformation, while the temperature is still high. It is a recovery process, starting with *nucleation* giving rise to *grain growth.*

Apparent dip : false dip The angle formed by a bedding or fault plane, with the horizontal strata, measured in random, vertical or perpendicular sections. The angle is always less then the *true dip*.

Appressed fold A fold whose limbs are almost parallel.

Aseismic ridge This ridge is a fragment of continental crust; it is so named to distinguish it from the seismically active mid-oceanic ridge.

Attitude The position of a structural surface relative to the horizontal ones. It has *strike and dip* measurements.

Augen : augen structures These structures consist of large lenticular mineral grains or mineral aggregates having the shape of an eye in cross sections, in contrast to the shapes of other minerals in the rock. *(Eyed structure; phacoidal structure).* In metamorphic rocks, such as gneisses and schists minerals like feldspar, quartz, or garnet are squeezed into elliptical or lensoidel structures resembling eyes *(augen)*, These are commonly enveloped by layers of mica or chlorite.

Austrain orogeny Stille (1930 - 1940) described this Phanerozoic shortlived orogeny in Europe.

Autochthon : authochthone A rock unit which remained at its site of origin, rooted to its basement. Authochonous rocks can be deformed.

Autochthonous : authigenic The term was used by Naumann (1858) for rock units which remained at the site of their formation.

Autogeosyncline : residual geosyncline A parageosyncline without an adjoining uplifted area, containing mostly carbonates (Kay, 1947).

Axial plane A planar surface parallel to axis of a fold (Fig. 8).

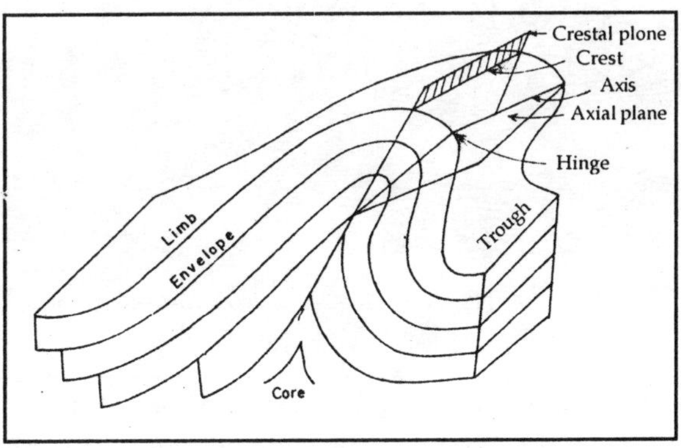

Fig. 8.

Axial plane cleavage : Which is parallel to the axial planes of folds. It can diverge on flanks of folds (*fan cleavage*). Sometimes it is parallel to the regional fold axes. (*slaty cleavage*). (Fig. 9).

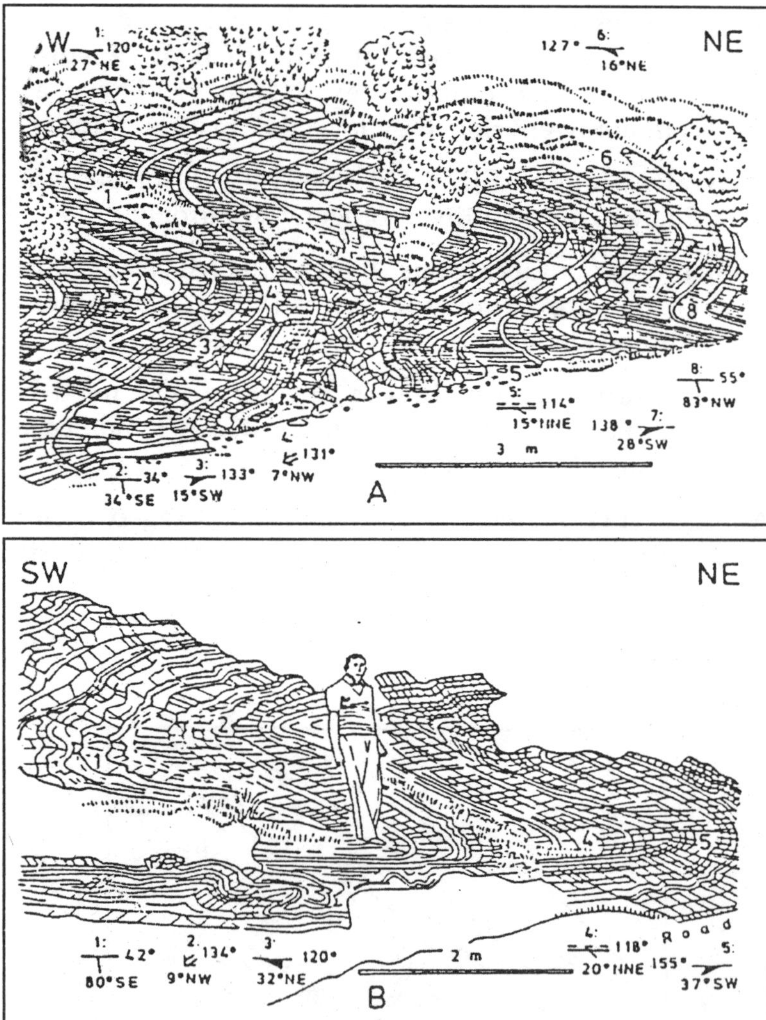

Fig. 9. (A+B). The beds and axes in recumbent F3 folds with fan cleavage - S3 in the Bhelunta slate-limestone within the Dunda Window, 4.8 km south of Dunda, Uttarakhand Area.

Axial symmetry : spheroidal symmetry In rock fabric having a unique axis of symmetry, and infinite number of mirror planes passing through that axis. (Fig. 10).

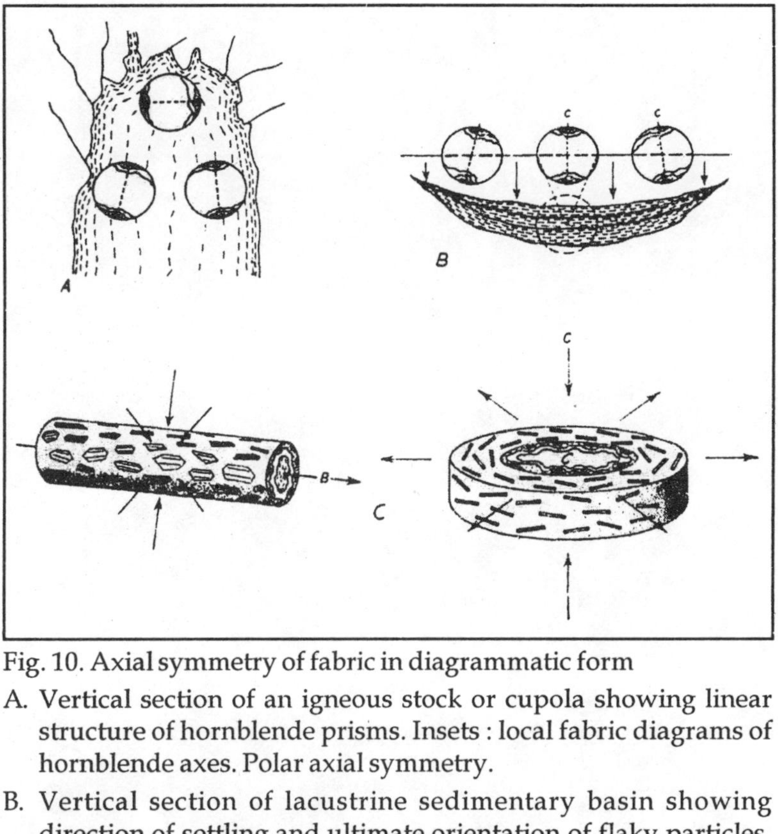

Fig. 10. Axial symmetry of fabric in diagrammatic form

A. Vertical section of an igneous stock or cupola showing linear structure of hornblende prisms. Insets : local fabric diagrams of hornblende axes. Polar axial symmetry.

B. Vertical section of lacustrine sedimentary basin showing direction of settling and ultimate orientation of flaky particles. Insets : local fabric diagrams of poles to clay flakes. Polar axial symmetry.

C. Central axial symmetry as expressed by the fabric of gypsum tablets in a rod produced by kneading (left) and of rutile needles in a disc obtained by 'spreading' of a ball of clay (right). Insets: integrated fabric diagrams (in Hills, 1973).

❑❑❑

B

Back-limb thrust fault : backlimb thrust fault A fault developed on the gently dipping "back limb" of an asymmetric anticline, in which the direction at the limb shows steeper angles (Douglas, 1950).

Back thrusting A thrust fault in an orogenic belt, showing the direction of displacement towards its interior, or contrary to the tectonic transport direction.

Baikalian orogeny An orogeny that occurred during the Precambrian-Cambrian transition, reported from the lower Riphean rocks. It is named after Lake Baikal in Russia.

Bald-headed anticline An eroded anticline whose crest is occupied by unconformable beds. A comon term used in petroleum geology.

Basement fold A compressional structure formed within a continent and affecting the entire thickness of the continental crust.

B axis It is one of the three orthogonal axes, *(a, b,* and *c)* and is used to describe the fabric geometry. This axis is at right angles to the unique plane of symmetry. In a deformation plan it has monoclinic symmetry. In simple shear the *b* axis lies in the shear plane at right angles to the direction of shear (See Fig. 1).

Bedding-plane cleavage : bedding cleavage; parallel cleavage It is parallel to the bedding plane.

Bedding -plane fault *bedding fault.*

Bedding -plane slip : flexural slip thrust The slip occurs of the sedimentary rocks along bedding plane during flexure folding (Fig. 11).

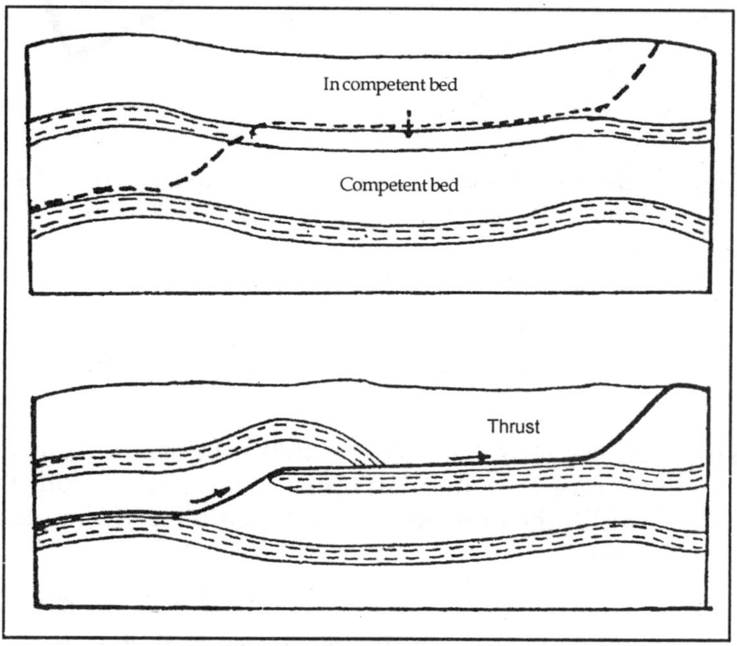

Fig. 11. Bedding plane slip thrust

Bedding thrust : *bedding glide.*

Benioff seismic zone A plane beneath the trenches of the Pacific belt, inclined toward the continents at an angle of about 45°, along earthquake foci. It is also called as the *Benioff fault zone.* The lithospheric plates sink into the mantle and cause earthquakes along this zone.

Beta diagram Plots of linear structures on a stereographic or equalarea net. Usually called as β diagram.

Block faulting A type of normal faulting in which the rockmass is divided into *faulted blocks* of different dimensions. (Fig. 12).

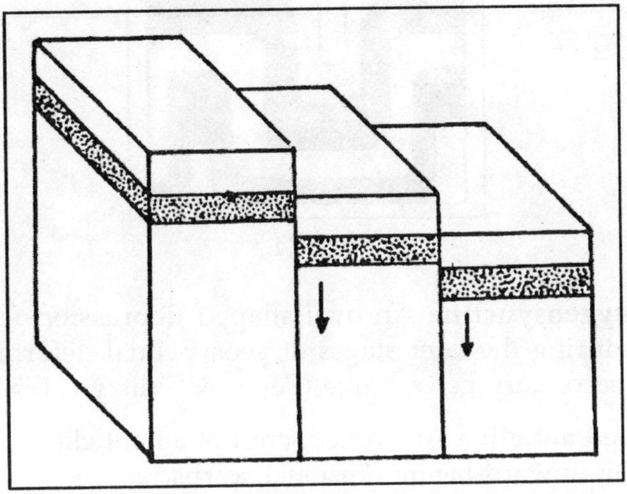

Fig. 12.

Boudinage : sausage structure This structure is developed in sedimentary and metamorphic rocks, in which a competent layer or bed between the less competent layers due to extension gets broken at regular intervals into *boudins* or sausages, which are elongated and parallel to the fold axes. (Fig. 13).

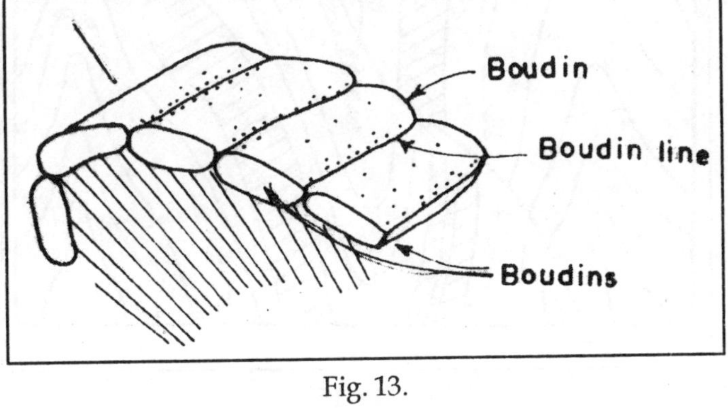

Fig. 13.

Box fold A fold style whose sides are rectangular (Fig. 14).

Fig. 14.

Brachygeosyncline An oval shaped depression formed during the later stages of geosynclinal deformation; secondary geosyncline (Peyve & Sinitzyn, 1950).

Brached anticline An eroded, crest of an anticline flanked by inward-facing erosional scarps.

Break thrust An overthrust formed due to deformation and fracturing of an anticline. (Fig. 15).

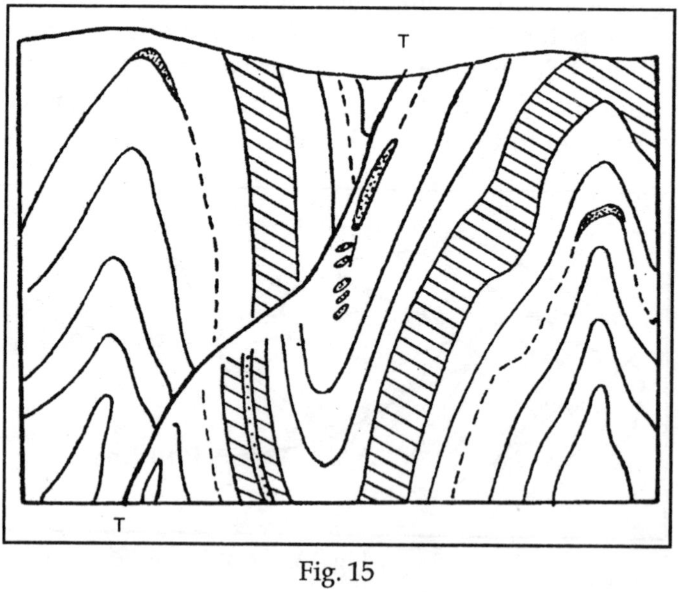

Fig. 15

Breccia (fault breccia, rubble rock) A coarse-grained rock, composed of angular broken rock fragments held together in a fine-grained matrix. Breccia may be formed due to tectonic processes

Brittle It is a rock-fracture produced due to less amount of deformation or strain.

Bysmalith : laccolith A vertical cylindrical igneous intrusion, bounded by faults. (Fig. 16).

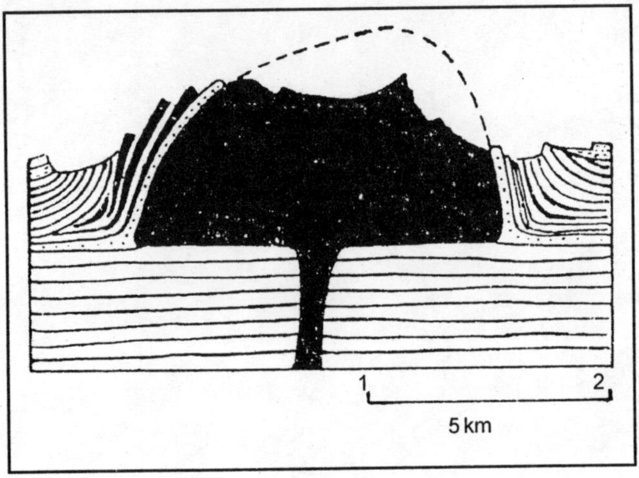

Fig. 16

C

Cactolith According to Hunt *et al.* (1953): "a quasi-horizontal chonolith composed of anastomosing ductiolith, or bulge, discordant like an akmolith or ethmolith". It has a cactus like form.

Cap rock Anhydrite-gypsum, calcite and sulphur bearing inpervious rock which, overlies the salt plug/salt dome.

Cariboo orogeny This orogeny occurred during early Palaeozoic time in the Cordillera of British Columbia.

C axis: *c* direction One of three orthogonal reference axes, *a, b* and *c,* that are used to describe the geometry of a fabric possessing monoclinic symmetry (Fig.1). In many *tectonites* the c axis is normal to the schistosity. It helps to describe a *deformation plan* that possesses monoclinic symmetry, caused by simple shear. The *c*-axis lies normal to the shear plane.

Chamoli earthquake Due to intraplate movement an earthquqake (Mag 6.8; IMD) occurred at Chamoli -Uttarakhand on 29th March, 1999. Its epicenter was between 30.4° N, 79.4° E with a focal depth of about 20 km. Rastogi (2000) opined that the rupture with intensity VI extended for about 250 x 150 km along NW-SE direction (Fig. 17a, b). Thrust faulting along ENE-WSW was the main cause of this earthquake.

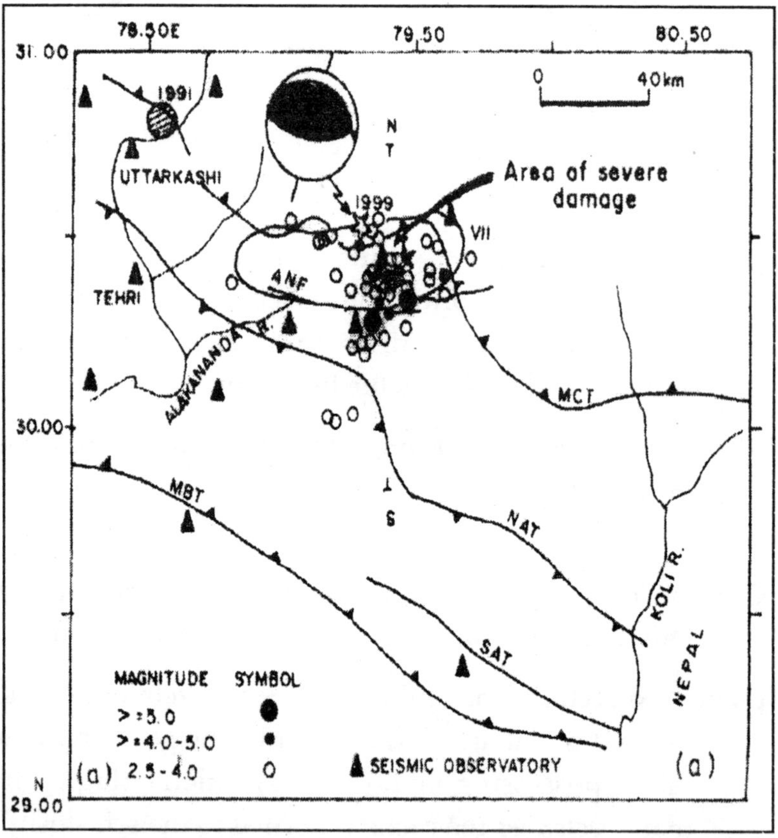

Fig. 17a: Epicentral map of Chamoli earthquake (1999) and its aftershocks (after Kayal, 2001). ANF - Alakananda fault, MBT - Main Boundary Thrust, MCT - Main Central Thrust - North Almora Thrust, SAT-South Almora Thrust.

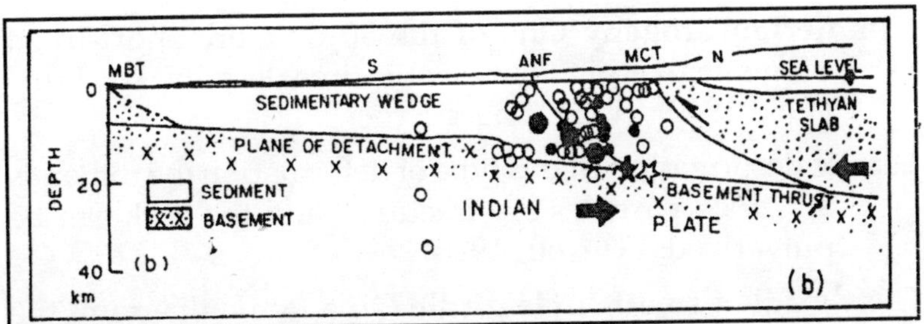

Fig. 17b: Model for the Chamoli earthquake and its after-shock, (after Kayal, 2001).

The aftershocks of this event were studied by IMD, NGRI (Rastogi, 2000) and G.S.I., (Kayal *et al.,* 2003). It may be remarked that the aftershocks were located towards north and south of the MCT. The model of Chamoli earthquake is similar to that of the Uttarkashi earthquake.

Chattermark *Marks* or scratches made on a rock surface by the faulted rock-mass.

Chevron fold A fold in which the hinge is angular. (Fig. 18).

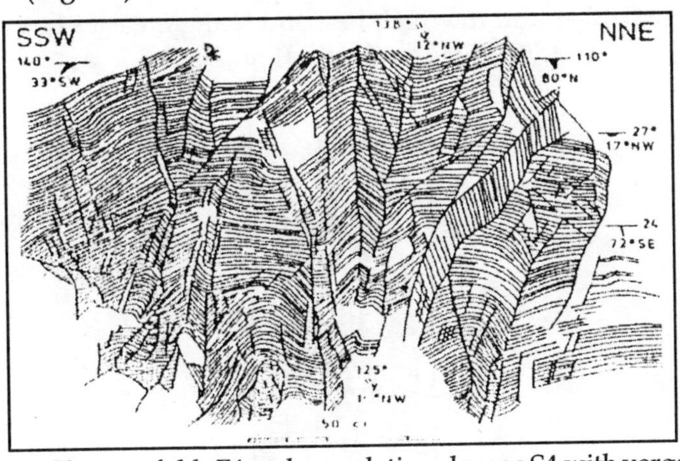

Fig. 18. Chevron folds F4 and crenulation clevage S4 with vergence towards SSW or NNE; Simla Slates exposed along the road slope at the exit NE of Jajal, Uttarakhand.

Cimmerian orogeny One of the 30 or more short-lived orogenies during Phanerozoic time described by Stille (1930-1940) in the Triassic rocks.

Clastic deformation In this type of deformation the minerals and rock particles are fractured and the rock can be pulverized. (Tyrrell, 1926).

Cleft girdle : small circle girdle On a fabric diagram, the annular maxima occupy the small circle of the net (Turner and Weiss, 1963). (Fig. 19).

Fig. 19. Monoclinic Symmetry with cleft girdie

Closed fold An old, term used for *isoclinal fold.*

Close fold A fold in which inter-limb angle varies between 30° and 70° (Fleuty, 1964). (Fig. 20).

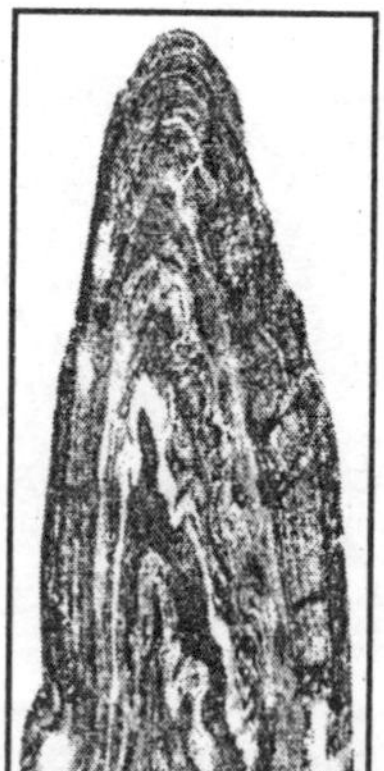

Fig. 20. Close folding of banded iron formation.

Closure : *structural closure.*

Collapse fault : It is a normal gravity fault on the margin of salt-domes in the Gulf Coast. Sediments overlying the salt dome collapse periodically due to salt - withdrawal.

Columnar jointing : Prismatic jointing, Parallel, Mural joints Polygonal parallel columns in basaltic rocks or in other extrusive and intrusive rocks. Due to cooling and contraction these joints are formed. (Fig. 21).

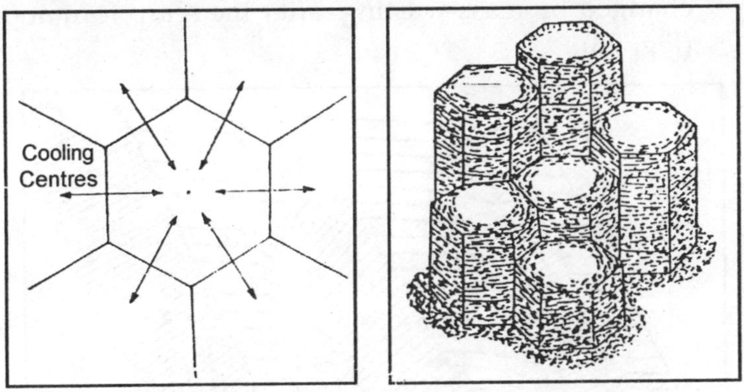

Fig. 21. Columnar joints (after Whitten and Brooks, 1972)

Composite unconformity An unconformity representing more than one event of nondeposition associated with erosion.

Compression A system of forces or stresses directed towards each other causing change of volume and shortening of substance. (Fig. 22A).

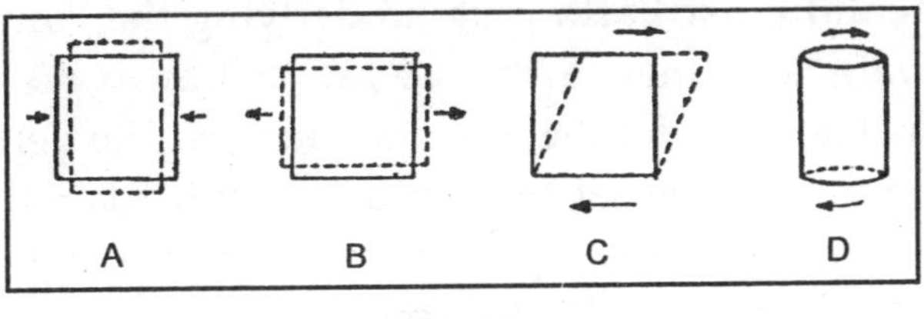

Fig. 22.

Conjugate [fault] The faults which are of the same age and deformational phase.

Conjugate [joint] Sets of joints formed due to deformation (usually compression).

Conrad layer : intermediate layer or the lower crustal layer The seismic zone of the Earth between the *Conrad discontinuity* and the Mohorovicic discontinuity.

Consequent fault scarp A fault scarp whose face is rapidly changed by mass-wasting after the scarp formation. (Fig. 23).

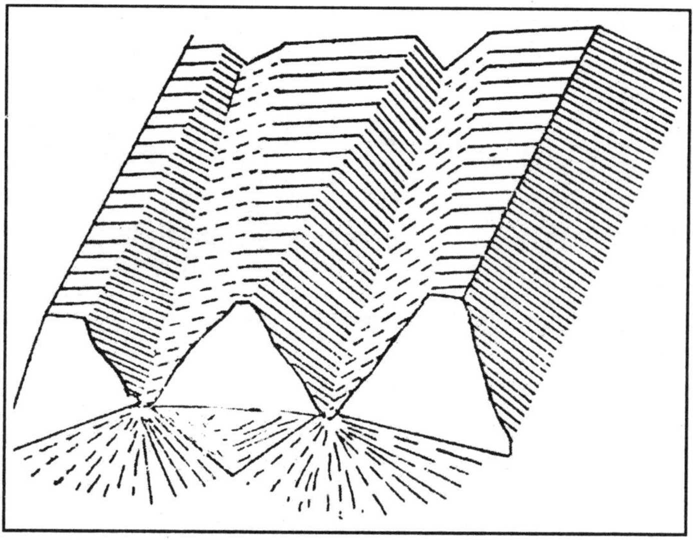

Fig. 23.

Continental crust Earth crust underlying the continents (*sial*), and measuring about 35 km to 60 km in thickness. under the mountain ranges. The density of the upper layer of the continental crust is ~ 2.7 and the velocities of compressional seismic waves through it are less than ~ 7.0 km/sec.

Continental displacement, displcasment migration. Wegener hypothesis; epeirophoresis theory; continental migration. A general term, propounded by Wegener (1912); also, called as *continental drift*. Wegener postulated the displacement of large continents (sialic) across a substratum of oceanic (simatic) crust, According to this theory the continents remained relatively fixed but the Earth expanded leaving wider gaps of oceanic areas. The continents moved away from each other by sea-floor spreading along a median ridge producing new oceanic areas beetween the continents. The masses away from the ridges consist of thick plates of continental and oceanic crust (Fig. 24).

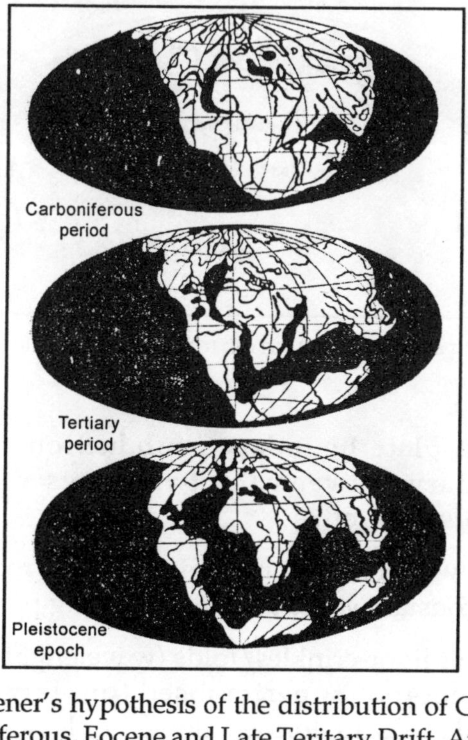

Fig. 24. Wegener's hypothesis of the distribution of Continents in Late Carboniferous, Eocene and Late Teritary Drift. Africa is in the centre.

Continental margin It is located between the shoreline and the deep ocean floor, including various rock provinces.

Convection A movement of subcrustal or mantle material, either laterally or in upward-or downward-directed *convection cells,* due to variation of geothermal heat. (Fig. 25).

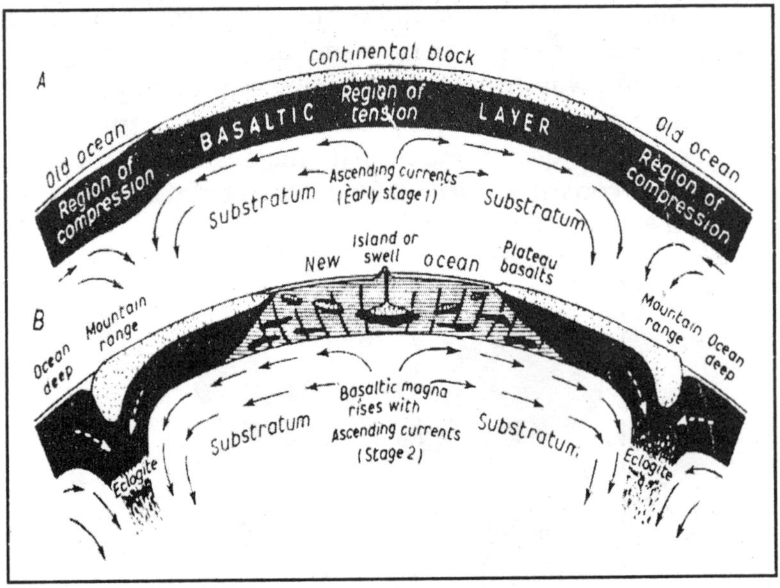

Fig. 25. Folding due to covection currents in the mantle *(From Holmes, 1965).*

Convergent plate boundary : subduction zone It is a boundary between the two plates which move towards each other.

Creep A slow, deformation (strain) of solid rock resulting from constant stress acting over a long period of time.

Crenulation Minor-crinkles/folds (wavelenght up to a few millimeters) superimposed on large folding. Crenulations are formed along the cleavage planes of a deformed rock.

Crenulation cleavage *(slip cleavage).* (Fig. 26).

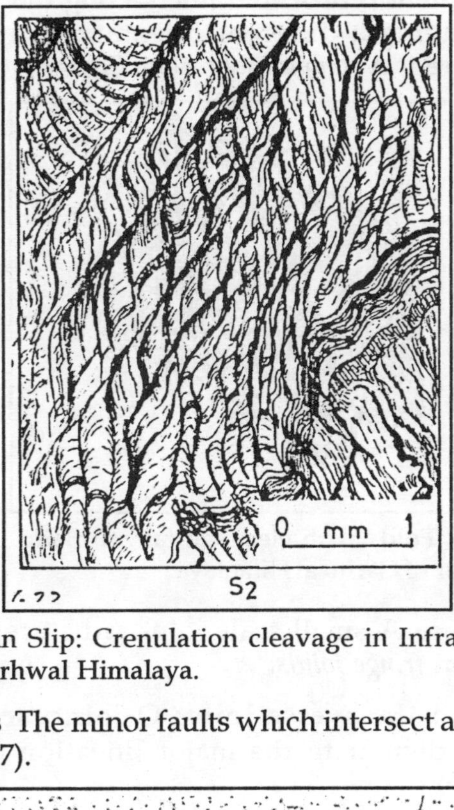

Fig. 26. Strain Slip: Crenulation cleavage in Infrakrol Slate in Rishikesh-Garhwal Himalaya.

Cross faults The minor faults which intersect a major fault. (Fig. 27).

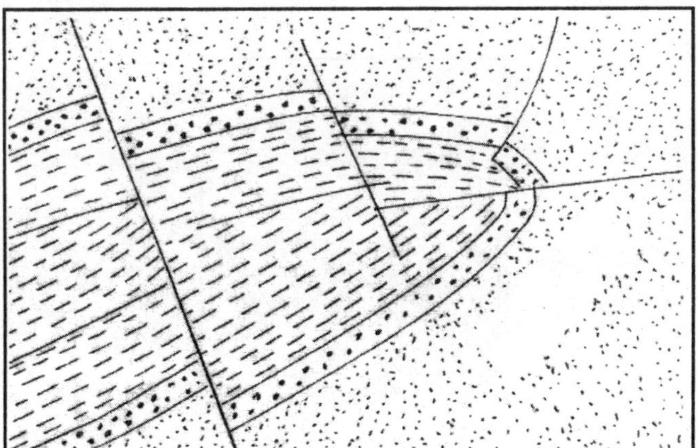

Fig. 27. Cross Longitudinal and Oblique faults

Cross fold : superimposed fold; transverse fold; subsequent fold A fold intersecting the pre-existing fold having a different orientation; the resulting structure is a *complex fold.* (Fig. 28).

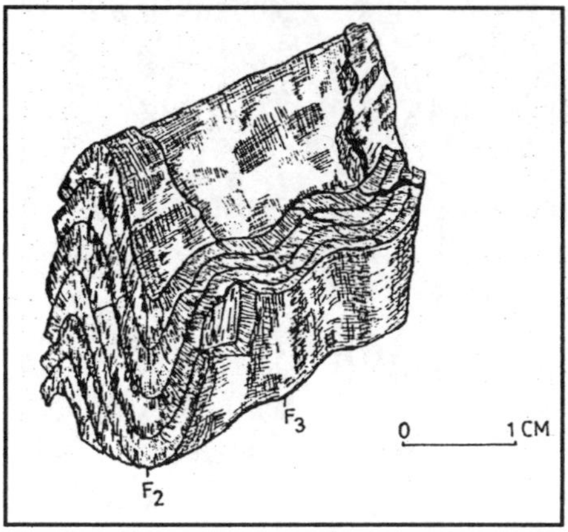

Fig. 28. Cross Folds with cleavage fan in Blaini Limestone at Narendranagar of Garhwal Himalaya.

Cross fracture A small-scale joint structure developed between *fringe joints.*

Cross joint : transverse joint; Q-joint; ac-joint It is perpendicular to the major lineation of the rock (Fig. 29).

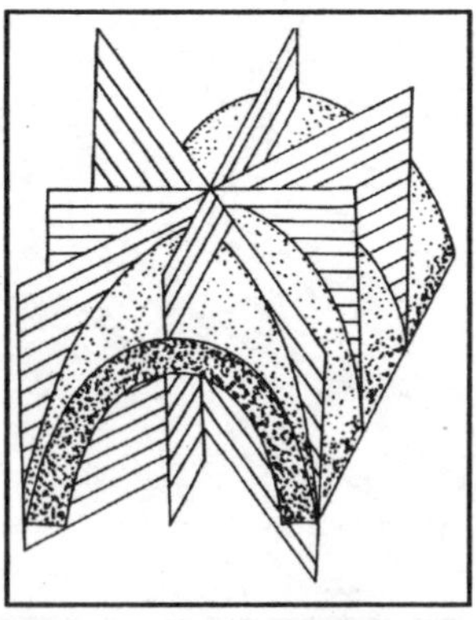

Fig. 29. Cross and Diagonal Joints

Crush breccia : cataclastic breccia A breccia formed by mechanical fragmentation of rocks during crustal movements. It is a *tectonic breccia* associated with the movements of folding or faulting. Norton (1917) used the term for a tectonic breccia in which the brecciation was accomplished without faulting or folding except for the rupture planes considered to be as minute faults.

Crystal gliding [cryst]; translation gliding; slip Deformation of crystalline material by orderly displacement of crystal structure Due to this crystal twins are formed. (Fig. 30-31)

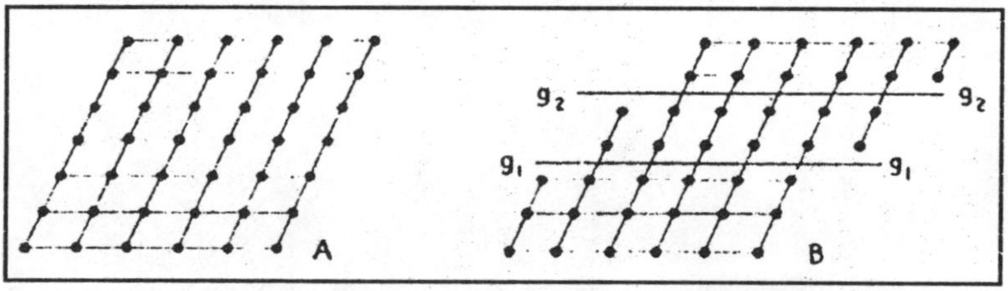

Fig. 30. Translation gliding, Each dot represents center of a unit of the space lattice. (A) Before gliding (B) After gliding along planes g_1g_1 and g_2g_2 (After Billings, 2000).

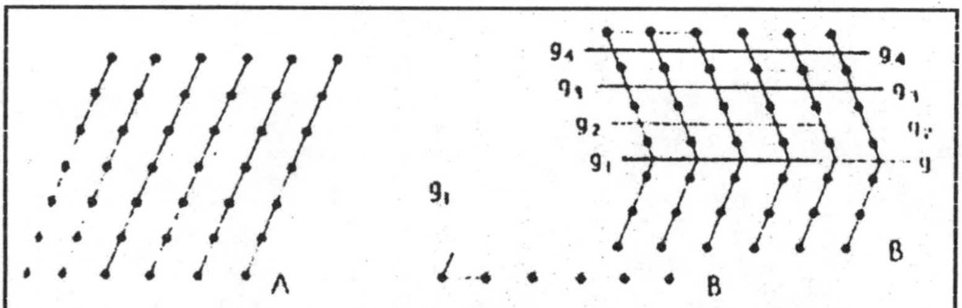

Fig. 31. Twin gliding, Each dot represents center of a unit of the space lattice. (A) Before gliding (B) After gliding along planes g_1g_1, g_2g_2, g_3g_3 and g_4g_4 (After Billings, 2000).

Crystal lattice : Bravais lattice; space lattice; direct lattice; translation lattice The three-dimensional regularly repeated set of points that represent a crystal structure. Each lattice has identical surrounding. There are fourteen partterns of lattice.

Crystal structure : crystalline structure The regular, and repeated arrangement of atoms in a crystal whose properties are described by the crystal lattice or space lattice.

□□□

D

Decollement Deformation, resulting in an independent style in the rocks above and below. It is associated with folding and overthrusting with formation of detachment structures of strata.

Decollement fold : superficial fold; Jura-type fold. In this folding the beds are independent of the basement, due to decollement.

Deformation A term used for the folding, faulting, shearing, compression, or extension processes of rocks due to the effects of tectonic forces.

Deformation fabric : tectonic fabric The fabric formed by deformation consisting of lineations, schistocities, cleavages, fold axes, and crystallographic preferred orientations.

Deformation lamella Planar features develoed due to deformation in the mineral grains (Fig. 32).

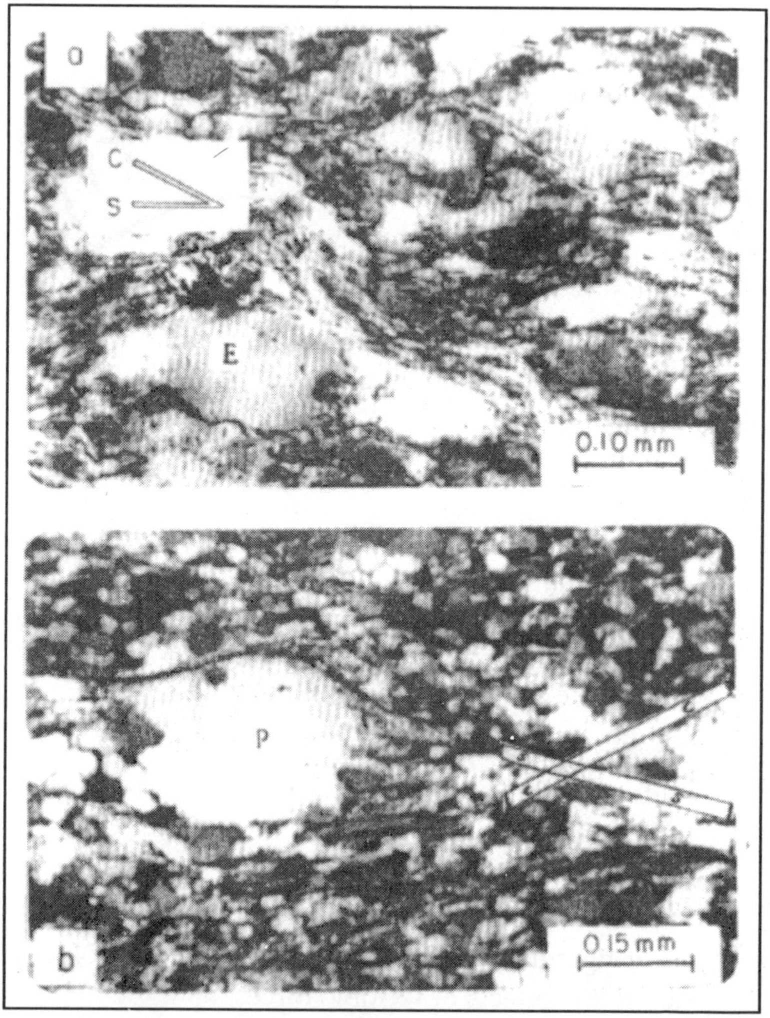

Fig. 32. a). Mylonitic foliations (S3), defined by preferred orientation of *deformation lamella* and C-represented by parallel alignment of micaceous minerals and recystallized quartz grains. (Crossed nicols PN2. b). Quartz porphyroclast (P) showing coinciding orientation of S and C surfaces and pressure shadows. indicating sinistral sense of shear. (Crossed Nicols. PN 1.4).

Diapir : piercement dome; diapiric fold; piercing fold. An anticlinal fold or dome in which the overlying rocks are ruptured by the squeezing-out of material of the plastic core. Igneous rocks can also occur as diapirs (Fig. 33).

Diapirism In this processs the rocks are anticlinally folded and ruptured. The geostatic pressure in the sediments can also form diapiric structures such as plugs. The process was first suggested for the formation of salt structures. (Fig. 33)

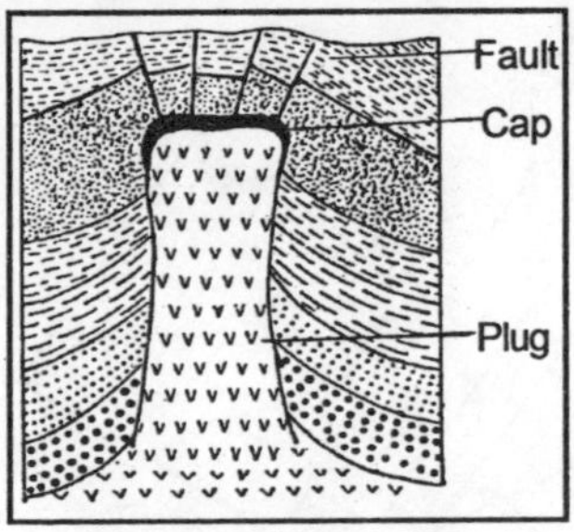

Fig. 33.

Diastrophism : tectonism A general term used for all movements of the crust formed by tectonic processes, inclusive of the ocean basins, continents, plateaus, and mountain ranges. *Orogeny* and *epeirogeny* are its major subdivisions.

Dike : dyke An igneous intrusion that cuts across the bedding or foliation of the country rocks. (Fig.34).

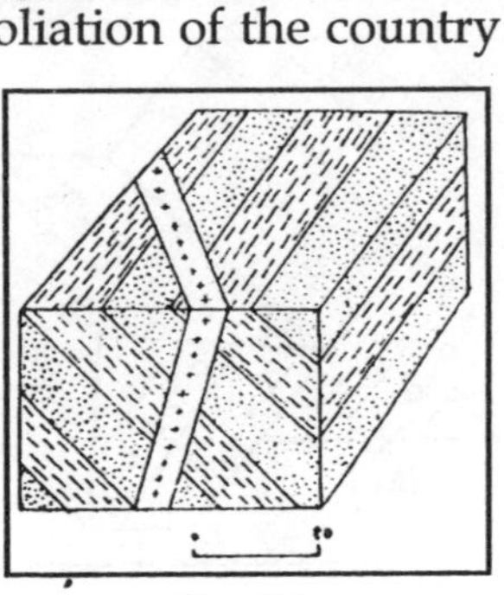

Fig. 34.

Dinarides The mountain range that stretches from northeastern Italy to Greece, paralleling the eastern Adriatic coast. It is the southwest-vergent branch of the Alpine mountain system.

Dip : true dip; angle of dip The angle that a structural surface, e.g. a bedding or fault plane, makes with the horizontal-level and is measured perpendicular to the *strike*. (Fig.35).

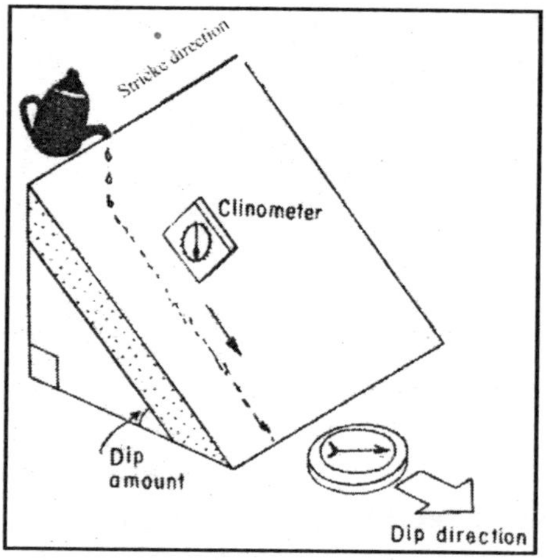

Fig. 35. Dip of a rock (after Lisle, 1988).

Dip fault A fault develped parallel to the dip of the strata. (Fig. 36)

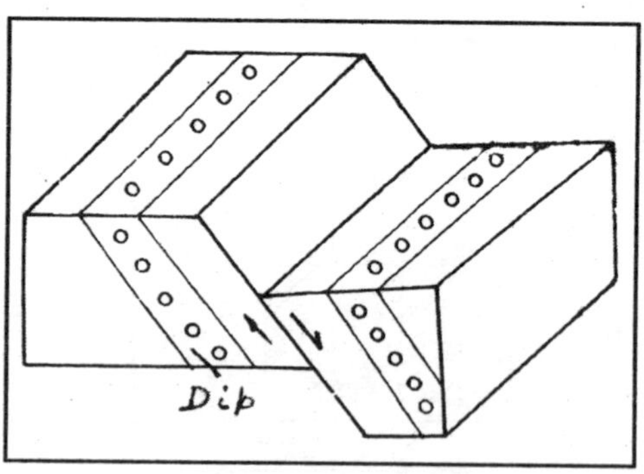

Fig. 36. Dip fault

Dip isogon A line joining points of equal dip. A classification of folds given by Ramsay (1967) which is based on dip isogons. (Fig. 37)

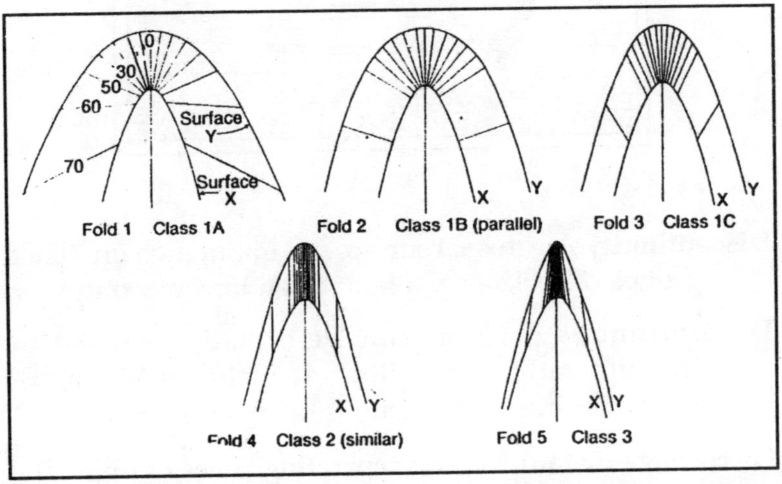

Fig. 37. Fundamental types of fold classes. Dip isogons at 10° intervals, from the lower to the upper surfaces X and Y. (After Ramsay, 1967)

Disconformity : parallel unconformity; erosional unconformity; nonangular unconformity; stratigraphic unconformity; paraunconformity. An *unconformity* in which the bedding planes above and below the break are essentially parallel, indicating a significant interruption in the orderly sequence of sedimentary rocks, characterized by an interval of erosion (or some times of nondeposition). It is usually marked by a visible and irregular or uneven erosion surface of relief (an unconformity) in which the older rocks remain horizontal during erosion or during simple vertical uplift and sinking of the crust (without tilting or faulting) The term applies to breaks in sedimentation elsewhere by rock units (Stokes & Varnes, 1955). The term formerly included *paraconformity* (Fig. 38).

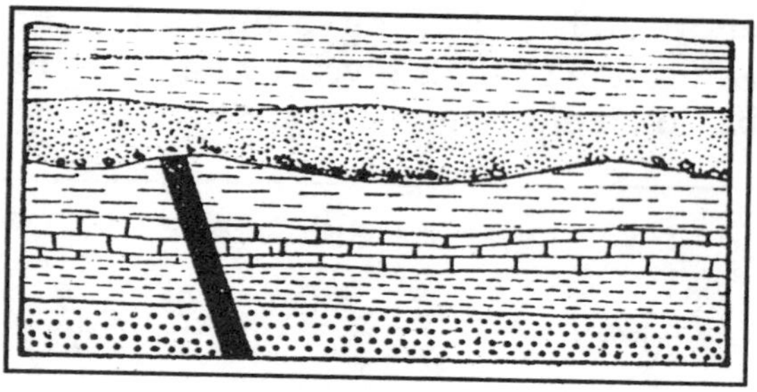

Fig. 38.

Discontinuity An erosnal surface separating two unrelated groups of rocks (e.g. a fault or an unconformity).

Discontinuous deformation Deformation caused by fracture rather than flow. It includes kinematic discontinuities, e.g. fractures or cleavage planes.

Discordogenic fault A fault separating zones of uplift and subsidence, which remained active during several geologic periods (Nikolaev, 1959).

Discrete A rock having a definite boundary with adjacent rocks.

Disharmonic fold A fold that varying in profile in the various layers through which it passes (Fig. 39).

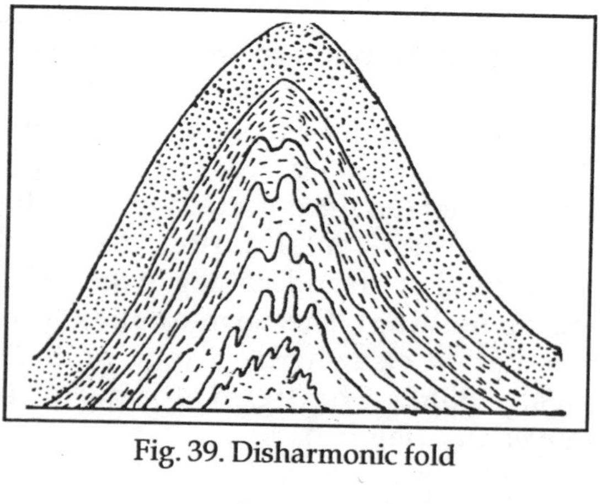

Fig. 39. Disharmonic fold

Displacement : dislocation A general term used for the relative movement of the two sides of a fault, measured in any chosen direction. (Fig. 40).

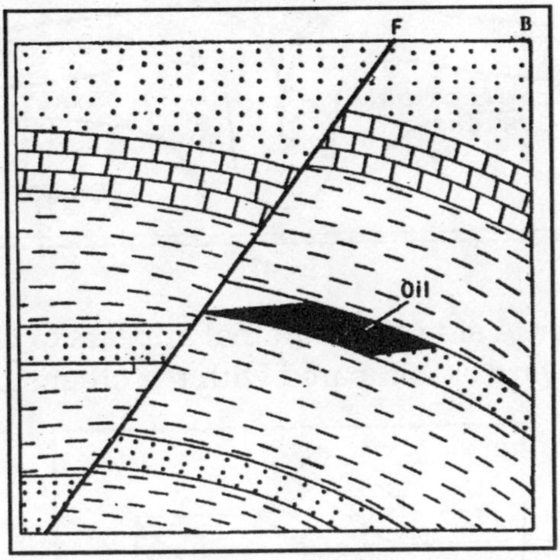

Fig. 40. Fault with displacement having an oil trap.

Domain (*fabric domain*).

Dome [fold] dome structure; structural dome; quaquaversal fold. An uplift or anticlinal structure, either circular or elliptical in outline, in which the rocks dip gently away in all directions. A dome may be small, and can extened for many kilomters in diameter. Domes include diapirs, volcanic domes, and cratonic uplifts. (Fig. 41).

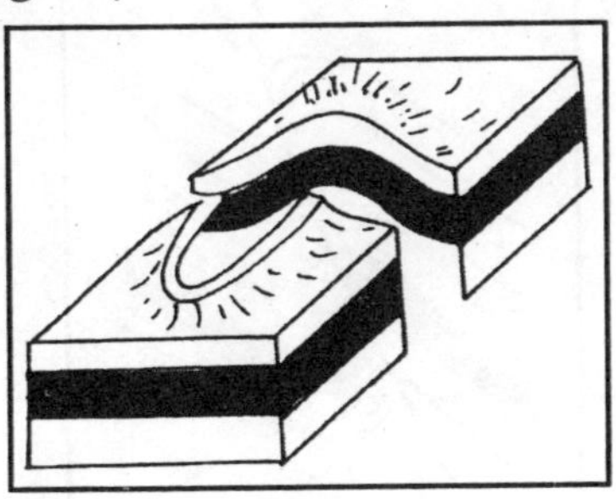

Fig. 41. Quaquaversal fold.

Downbuckle : tectogene Downfolding of sialic crust, associated with oceanic trenches. (Fig.42).

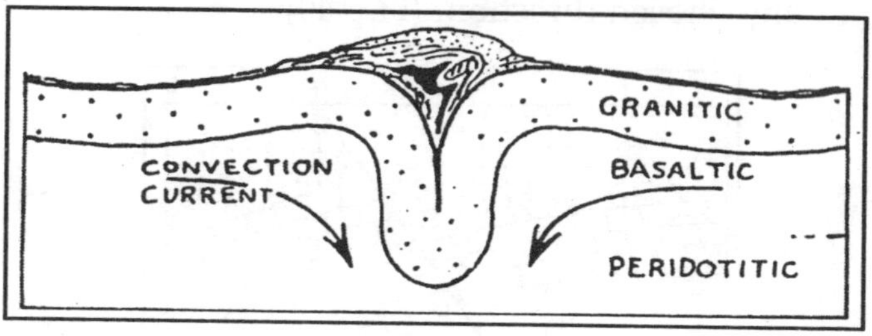

Fig. 42.

Downthrown A side of a fault that appears to have moved downward as compared with the other side (Fig. 43).

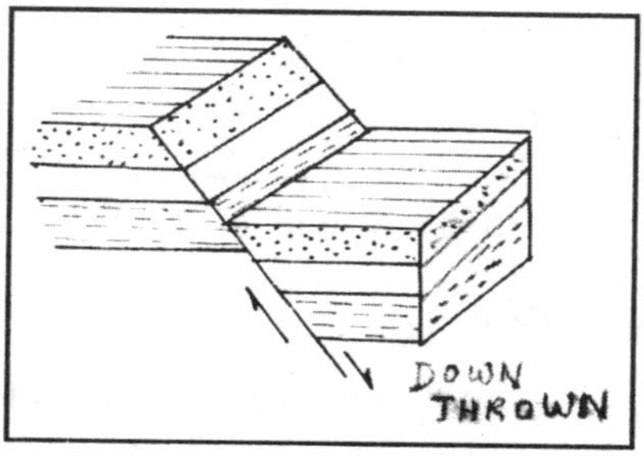

Fig. 43.

Drag The bend or distortion of starta on either side of a fault, caused by the friction of the moving blocks along the fault surface. (Fig. 44).

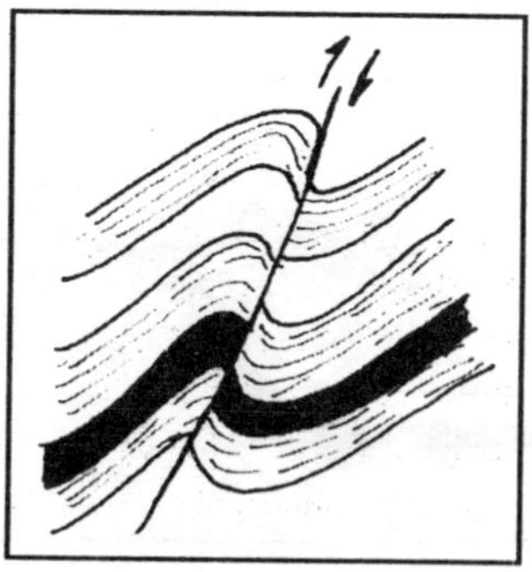

Fig. 44. Drag.

Drape fold (a) A *supratenuous fold or compaction fold.* (b) A fold produced in layered rocks by movement of an underlying brittle block at hight angles to the layering (Friedman *et al.,* 1976). It is a type of a forced fold which excludes supratenuous or compaction folds (Fig. 45).

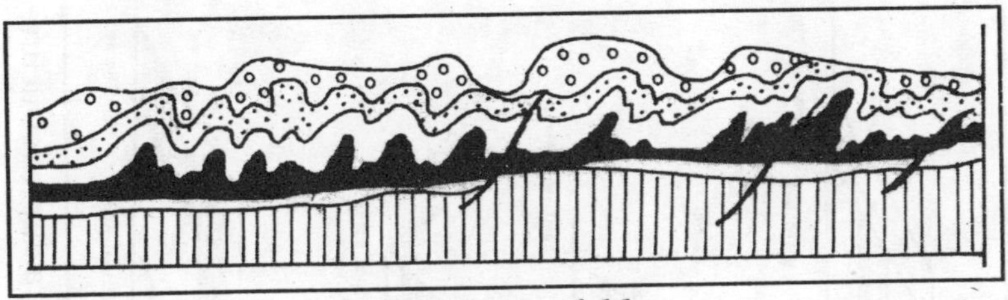

Fig. 45. Drape folds.

Duplex The new or younger thrusts are developed in the footwall of the older thrust due to progressive failure of footwall ramps and their abandonment (Norris, 1958. Bahuguna and Saklani, 1988; Boyer and Elliot, 198; Butler, 1982; Cooper, 1981, Dahlstrom, 1969, 1970). The bounded surfaces of the asymptotic faults can be rejoined with the earlier thrusts. The higher thrust forms the roof and the last sharing with the roof is the floor thrust. This structure consisting of imbricated/tectonic horses is collectively known as a Duplex. (Fig. 46). In the Bhagirathi valley the Vaikrita is the roof while the Budhahder is the floor thrust and the duplex is formed by the Barsu, Bhatwari and the Sainj horses (Saklani, 1993). Similarly in the Yamuna valley region, the MCT, IIa,b and the MCT III constitue a major duplex (Figs. 46 and 47).

contd... fig. 46 & 47

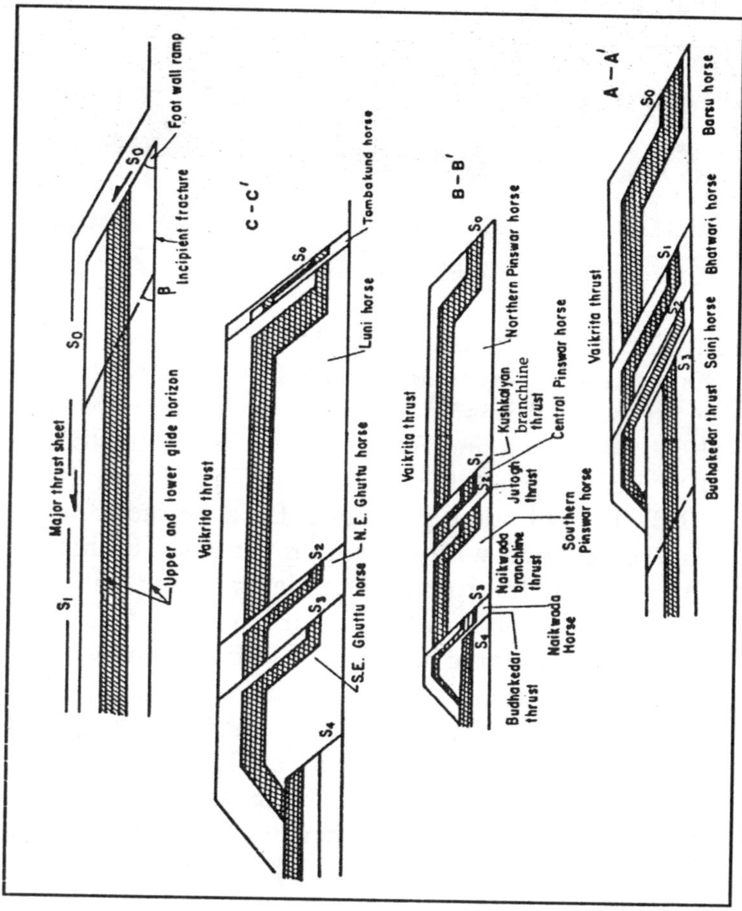

Fig. 46. Progressive collapse of footwall ramp builds up a duplex structure.

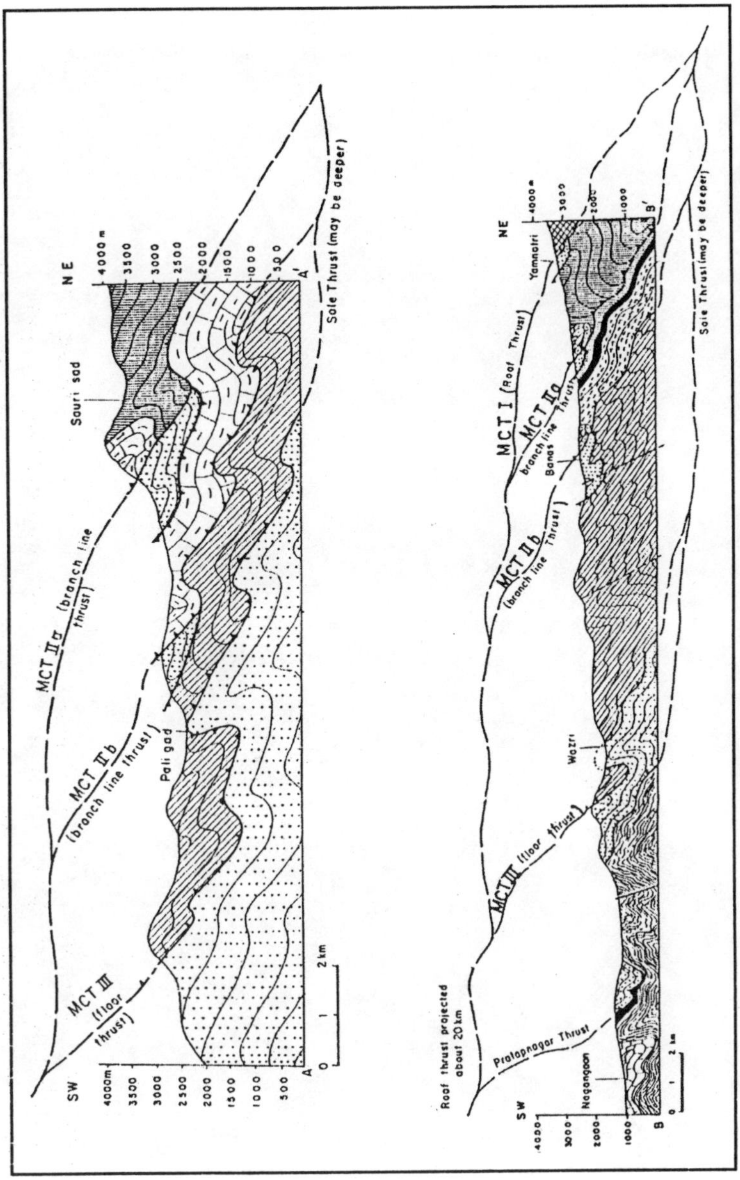

Fig. 47. Geometry of the composite Main central thrust (MCT) in the Yamuna Valley, Garhwal Himalaya.

E

Elastic deformation Deformation of a mass which disappears due to removal of forces. Commonly, this type of deformation follows the *Hooke's law.*

Elasticoplastic A deformation that has elastic and plastic phases.

Elastic strain The elastic behaviour of a material during deformation. (Fig. 48).

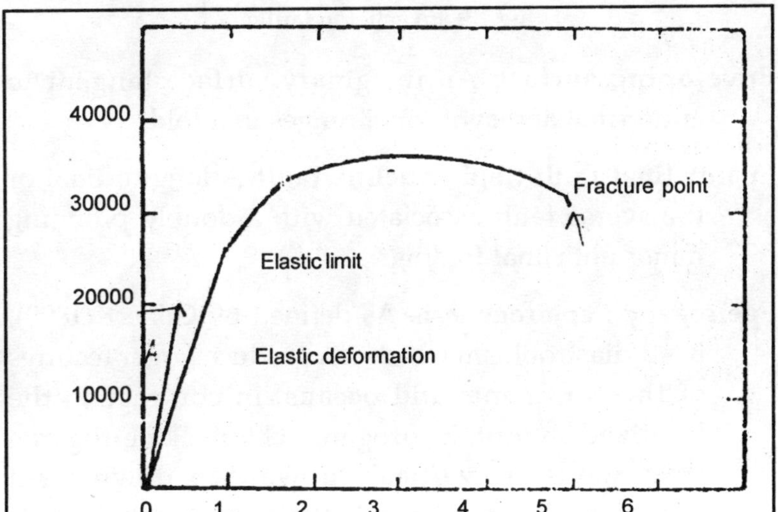

Fig. 48. Elastic strain curve

En echelon : steplike arrangement Structures with staggered arrangement (faults). In this zone , the strike of the individual fractures is oblique to that of the fault zone. (Fig. 49).

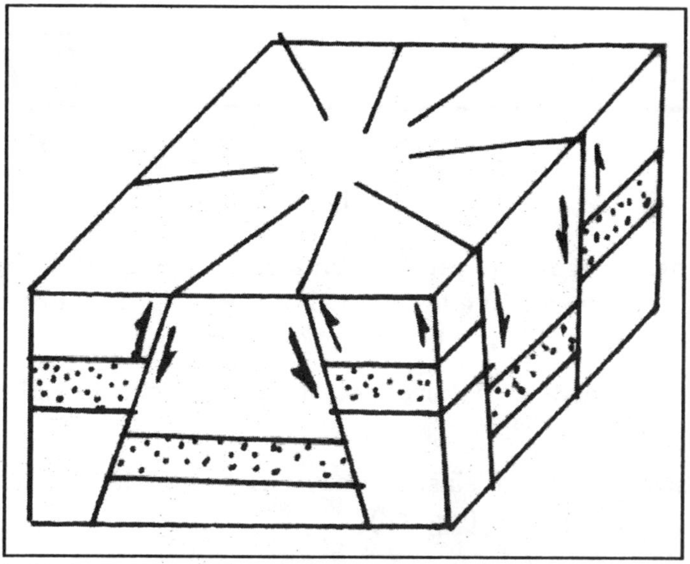

Fig. 49. En echelon faults

Enveloping surface An imaginary surface tangent to antiformal and synformal hinges in a fold

Epanticlinal fault : epi-anticline fault A longitudinal or transverse fault associated with a doubly plunging minor anticlinal folding.

Epeirogeny : epeirogenesis As defined by Gilbert (1890), it is a diastrophism which can produce larger features of the continents and oceans, in contrast to the localized mountain orogenic chains. Epeirogenic movements are vertical, (upward or downward) and affect the continents, in the cratons and orogenic belts.

Epicenter : epicentrum. The point located directly above the focus of an earthquake. (Fig. 50).

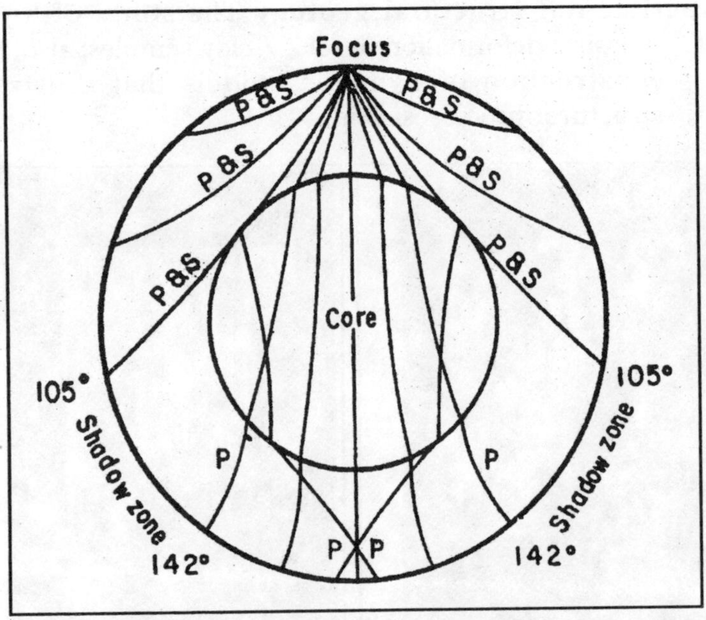

Fig. 50. Elastic waves through the earth (after Whitten and Brooks, 1972).

Erian orogeny : Hibernian orogeny One of the 30 or more short-lived orogenies during Phanerozic time.

Erosional unconformity An unconformity formed by erosion. It separates older rocks from the younger rocks that cover them.

Erosion thrust A type of thrust fault in which the hanging-wall block moves across an erosional surface.

Euler's theorem In it any displacement of a spherical surface over itself leaves one point fixed. It is used in *plate tectonics.* According to this the displacement of a rigid body along an axis may be considered as a rotation axis through a point on the sphere.

Exhumation Exposure of a buried geological feature exposed due to erosion.

Experimental structural geology The study of high-pressure deformation of rock / clay samples; also, the construction of dynamic models that illustrate structural processes. (Fig. 51).

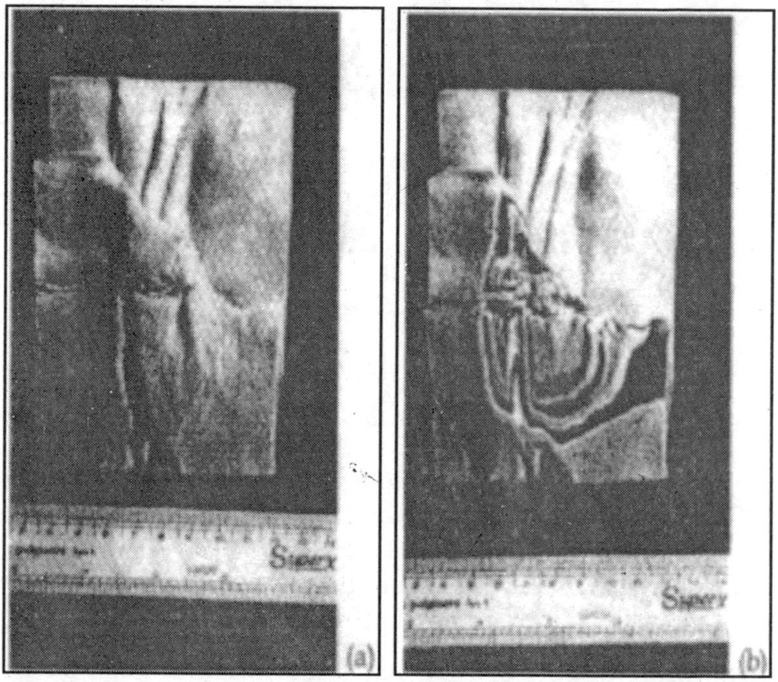

Fig. 51. a). Devolpment of early folds (fold hinge lines trending left to right) and superposed folds (trending top to bottom on a deformed layer surface (after 25% early shortening and 23% superposed shortening). b). Removal of a horizontal slice from the upper part of the model revealing partial klippen (Dubey and Jayangondaperumal, 2005).

Extension fault A fault in sedimentary rocks along which there has been bed-parallel elongation giving rise to tectonic thinning (Norris, 1958, 1960).

Extension fracture A fracture that develops perpendicular to the direction of greatest stress and parallel to the direction of compression; a *tension fracture.* See also: *extension joint; tension crack.* (Fig. 52)

Fig. 52. Extension fracture

❑❑❑

F

Fabric Mutual relationship of components of a deformed rock including texture, structure, and preferred orientation of minerals. It describes the shapes and characters of a rock mass and the orientation pattern in space. The term was first used by Sander (1930).

Fabric analysis Analysis of the geometrical parts of the rock fabric.

Fabric axis : reference axis; tectonic axis One of three orthogonal axes representing orientation of fabric elements, and movement symmetry of deformed rocks.

Fabric digram : petrofabric diagram The stereographic or equal area projection of fabric elements of a rock. (Bhatt and Saklani, 1990, Fig. 53).

Fabric domain It defines structural boundaries or compositional discontinuities in a rock, more or less homogeneous in mineral fabric.

Fabric element A component of a rock fabric.

Face A term used by Shrock (1948) for the original top or upper surface of a rock strata. Commonly the face of the fold is towards the stratigraphically younger rocks. An upright fold faces upward while an overturned anticline faces downward. Faults face towards the structurally lower units.

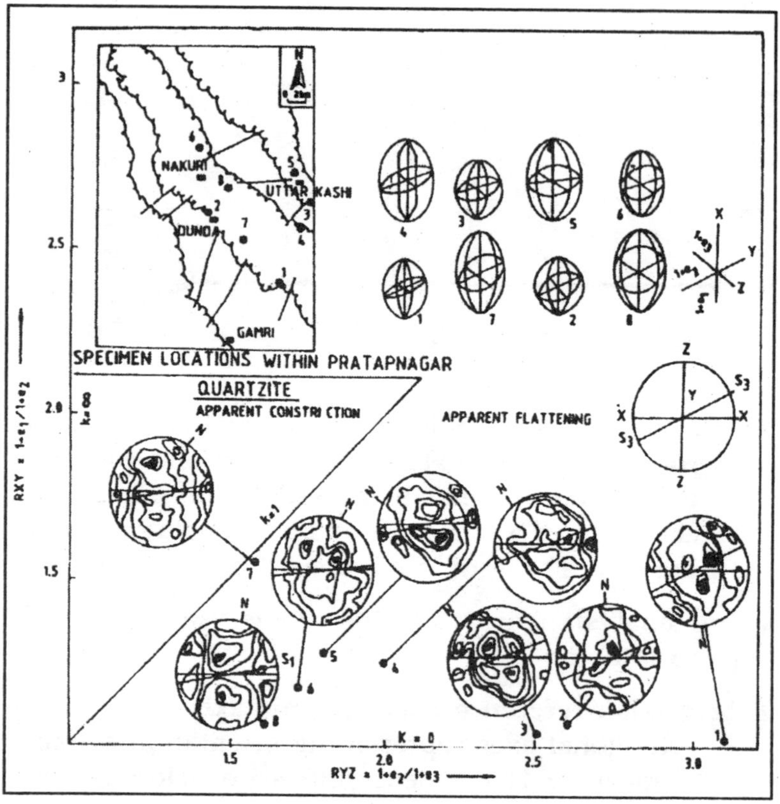

Fig. 53. C-axes fabrics and estimated strain of host old quartz grains in Pratapnagar Thrust Sheet. The inset geotectonic map shows the specimen locations of Pratapnagar Quartzite. All the spectimens are in the apparent flattening field showing oblate shape of strain ellipsoids. (Bhatt and Saklani, 1990).

False folding Folds which are not genetically related to lateral compression, e.g., shear and supratenuous folds.

Fan cleavage : cleavage fan Cleavage planes along the axial planes occur at low to large angles along the limbs of folds with fan shaped orientation. (Fig. 54). Schwan (1980) reported the occurrence of this cleavage from Narendranagar area of Garhwal Himalaya.

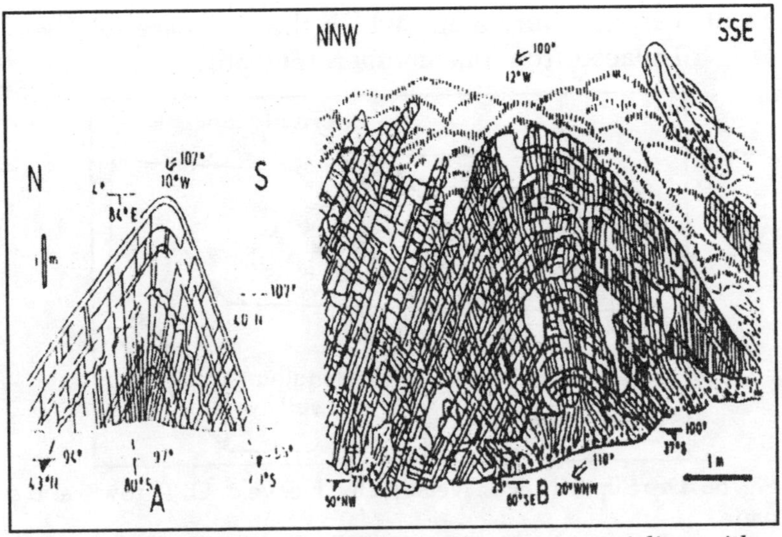

Fig. 54. (a) Inclined and weakly south-vergent anticline with a distinct cleavage fan (F_3 + S_3 types). (b) Inclined north-vergent anticline with fracture cleavage in the fan (F_3 + S_3 types). Quartzite beds of Infra-Krol exposed at the southern exit of Narendranagar, Garhwal Himalaya. (After Schwan, 1991).

Fan fold A fan shaped fold with a broad *hinge* in which the limbs appear converging away from the hinge area. (Fig. 55).

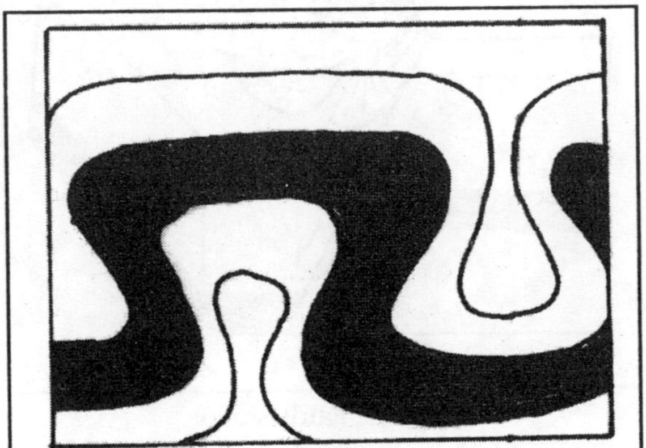

Fig.55.

Fault It is a fracture along which the rocks are relatively displaced from one another. (Fig. 56).

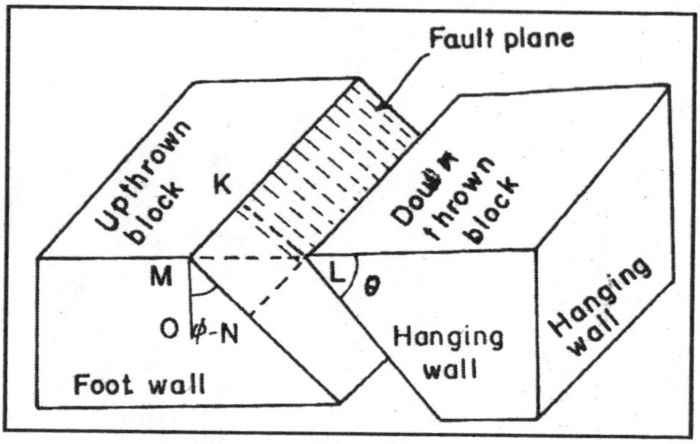

Fig. 56. Dip Slip MN, ML, Netslip ON, Heave or O, Throw θ and ϕ hade

Fault apron Deposition of a a rock waste along the base of a fault scarp characterized by numerous alluvial cones. (Fig. 57).

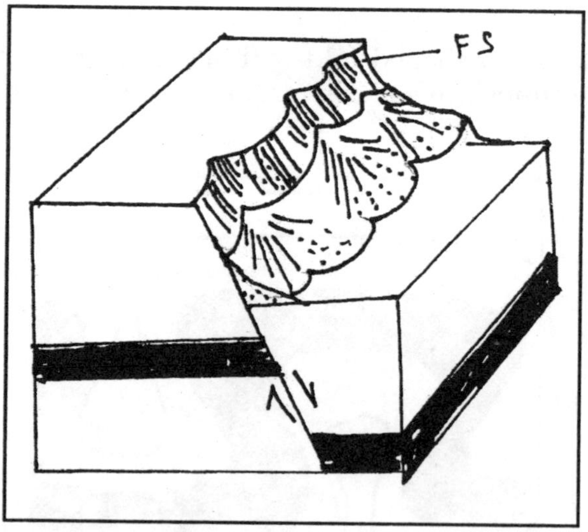

Fig. 57. Fault Scarp

Fault bench (*fault terrace*).

Fault block A rock unit boundary faults due to block faulting. (Fig. 58).

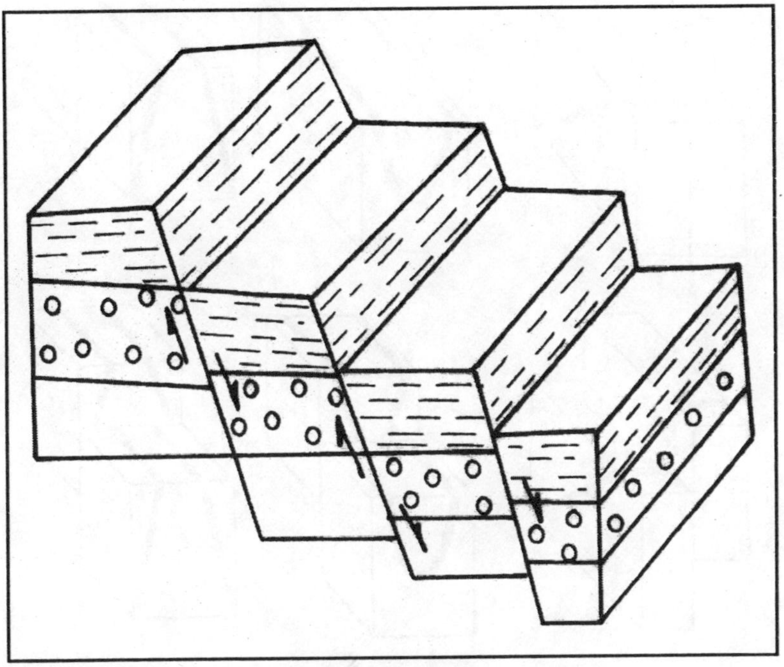

Fig. 58. Fault Blocks

Fault-block mountain *(block mountain).*

Fault breccia : dislocation breccia A tectonic breccia composed of angular fragments resulting from the crushing, shattering, or shearing of rocks during movement / and friction between the walls of the fault, or from ruptures associated with a major fault (a *friction breccia*). It is distinguished by cross-cutting relations, *fault gouge,* and by slickensided blocks.

Fault Effects When rocks are affected by faulting then these are characterised by repetition, omission etc. (Fig. 59).

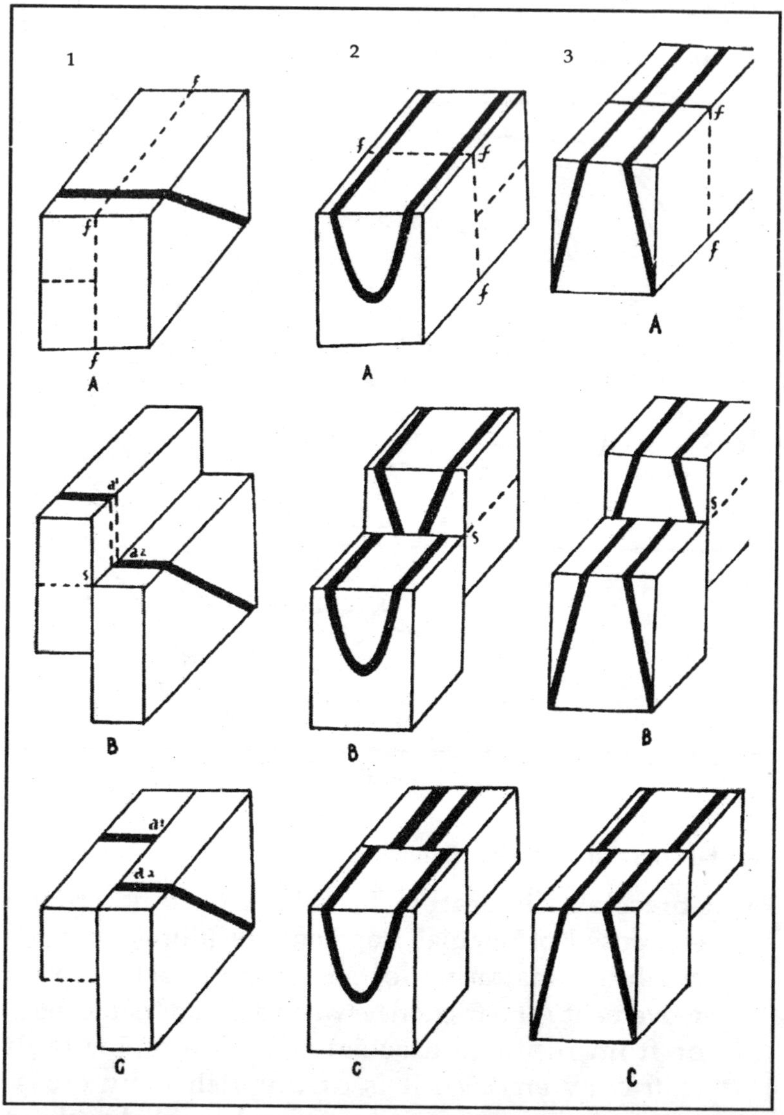

Fig. 59. Effects on outcrops produced by strike (1) and dip (2, 3) faults, show the effects of strike faults on topography. The outcrops of the shaded bed show repetition (c) 2 and 3 show the effects by dip faults on synclinal and anticlinal strata. The result is that outcrops have come closer in 2 and gone wide apart in 3 on the denuded upthrow side (cc)

Fault gouge : clay gouge; selvage Highly crushed claylike material, of minerals found along some faults or between the fault walls occupying a fault zone; at times, associated with *fault breccia.*

Faulting The structural process which produces *faults formed due to fracturing and displacement.*

Fault ledge *fault scarp.*

Fault plane A planar faulted surface.

Fault trace *(fault line).*

Fault wedge A wedge-shaped rock mass bounded by faults. (Fig. 60).

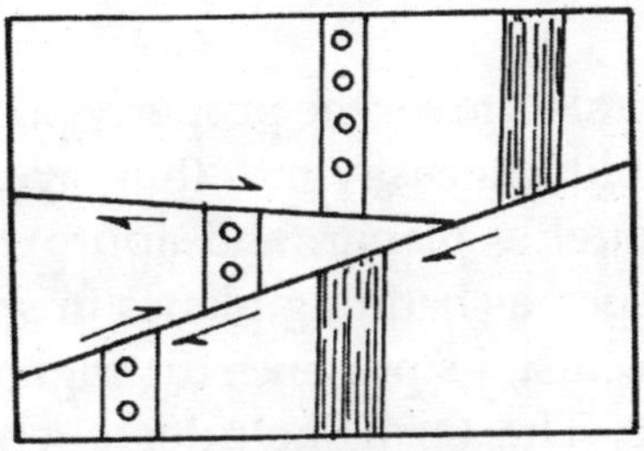

Fig. 60. Fault wedge

Feather jointing : pinnate jointing. A feather like joint pattern developed along a fault zone due to shear and tension. (Fig. 61). Saklani (1993) described the occurrence of pinnate joints in the Bhelunta locality of the Pratapnagar quartzite, Garhwal Himalaya.

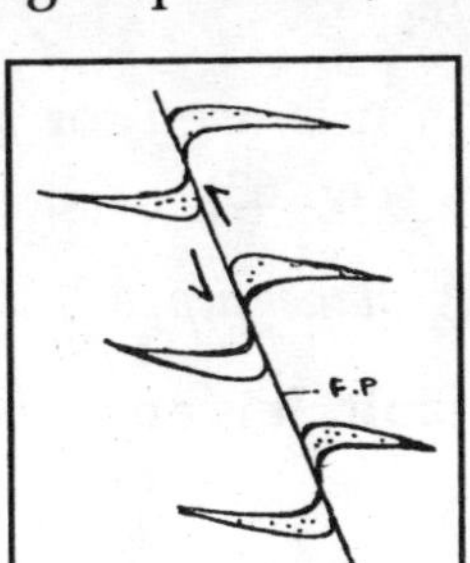

Fig. 61. Feather (pinnate jointing)

Fenster *(window)*. (Fig. 62).

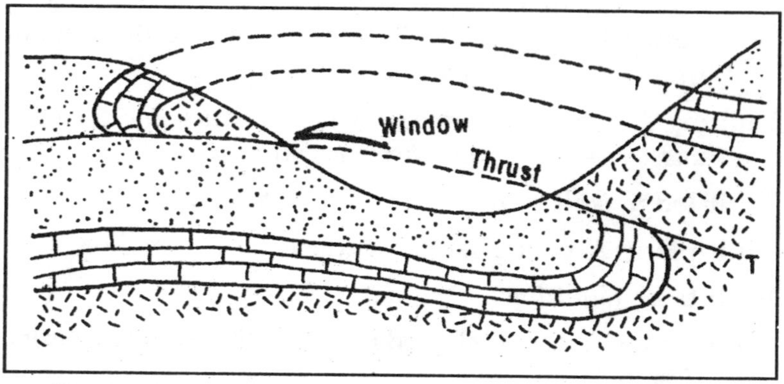

Fig. 62. Tectonic window (after Whitten and Brooks, 1972)

Fissility A general term for the property possessed by some rocks of splitting easily into thin layers along closely spaced, roughly planar, and approximately parallel surfaces, such as bedding planes in shale or cleavage planes in schist; its presence distinguishes shale from mudstone. The term includes such pheomena as *bedding fissility* and *fracture cleavage.*

Flank *(limb).*

Flat joint In igneous rock, joint dipping at an angle of 45° or less.

Flattening *(ellipticity).*

Flexural fold This term is used for *flexure-flow folds* and *flexure-slip folds.* (Fig. 63).

Flexural slip *(bedding-plane silp)*

Flexural-slip thrust fault It is an *uplimb thrust fault.* (Price 1965).

Flow joint A joint developed parallel to the flow layers of a plutonic igneous rock (Tomkeieff, 1943).

Fold A curved or bent planar structure of bedding plane, foliation, or cleavage. formed due to deformation.

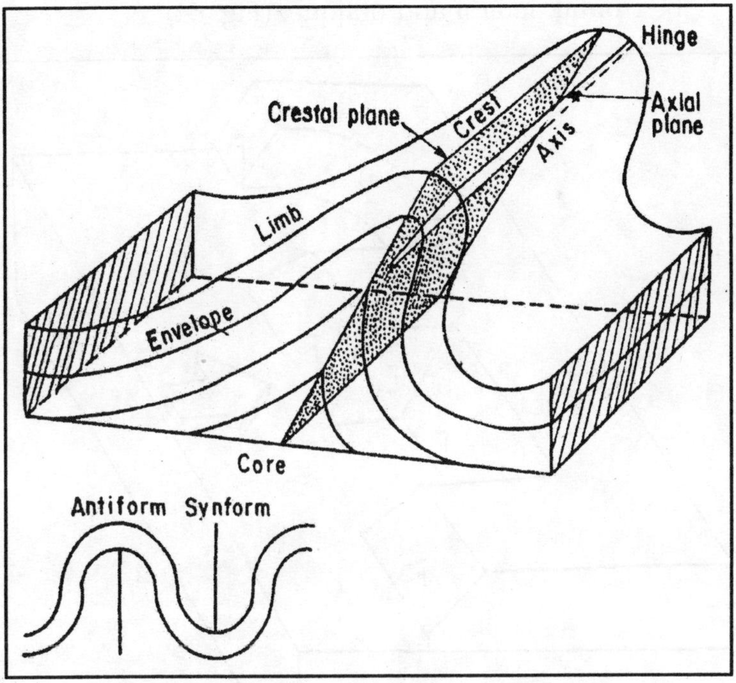

Fig. 63. Anticlinal flexure-fold and its parts

Fold breccia *(tectonic breccia)* It is made by angular fragments and brittle rock layers.

Folded fault The fault which is deformed by folding. *Step fault* can also fold the hangingwall beneath it (Jones, 1971).

Fold facing The middle limb of an overfold replaced by a fault surface.

Fold mountains Mountains formed due to largescale folding and subsequent uplift of rock strata.

Fold mullion The cylindrical undulations of bedding giving rise to a type of mullion structure.

Footwall Ramps and Flats : Ramps and flats developed towards the footwall side of a thrust sheet. These can be frontal, lateral and oblique. (Fig. 64).

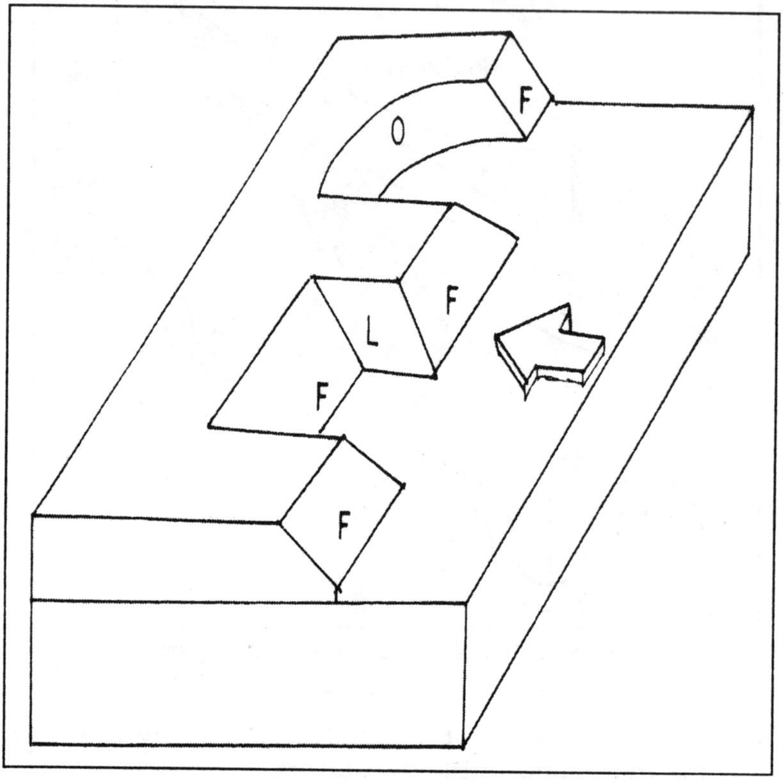

Fig. 64. Footwall side of a thrust sheet F- Frontal L-Lateral O-oblique ramps.

Forced fold : forced folding : drape fold. A fold whose style and trend are dominated by the shape of some forcing member below.

Forelimb The steeper limb of an asymmetrical anticlinal fold.

Fracture A break in a rock, produced due to mechanical failure by stress. It includes cracks and joints.

Fracture cleavage : close-joints cleavage A type of rock cleavage in deformed and metamorphosed rocks reprsented by closely spaced, parallel joints and fractures. (Fig. 65)

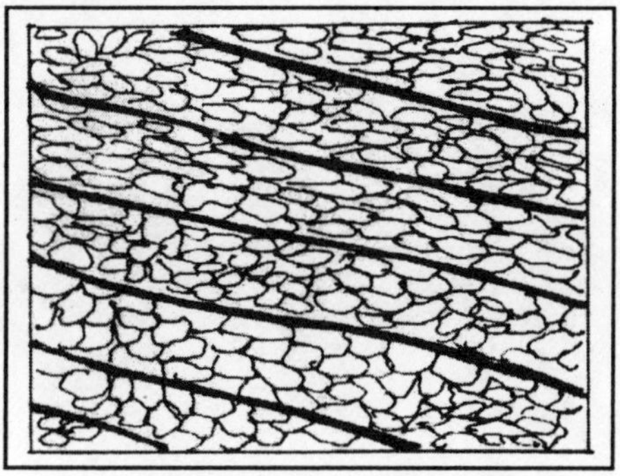

Fig. 65. Fracture cleavage

Front-limb thrust fault : forelimb thrust A thrust fault formed on the "front limb" of an asymmetric anticline. (Fig.66). The tectonic transport direction is toward the adjacent syncline ; the fault dips in the same direction as the anticlinal axis, elongating the limb of the anticline (Douglas, 1950).

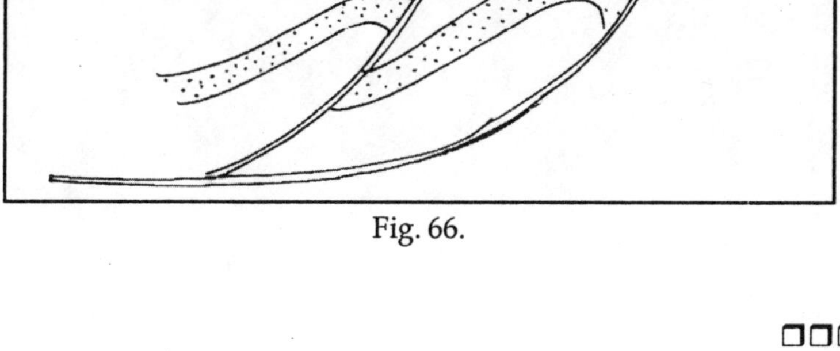

Fig. 66.

G

Geodynamics This branch deals with the tectonic forces and processes of the Earth's interior.

Geosyncline A slowly sinking downwarping either elongate or basin like for several of kilometers, in which sedimentary and volcanic rocks accumulate to thicknesses of thousands of meters. A geosyncline may form in part of a tectonic cycle (Fig. 67). The theory was prpounded by Hall (1843) and the term *geosynclinal* was given by Dana (1873). Different opinions about the origin, mechanics, and essential features of geosynclines are reflected in the literature. All geosynclinal phenomena are related to opening and closing of oceans (Wilson, 1968).

Geotumour According to Haarmann (1930), it is a regional uplift.

Gridle A concentration of points represnting fabric elements on an equal-area projection. When concetration is seen along the great circle then it is referred to a great circle girdle and if it is located on the small circle then it is called a small circle/cleft girdle (Turner and Weiss, 1963).

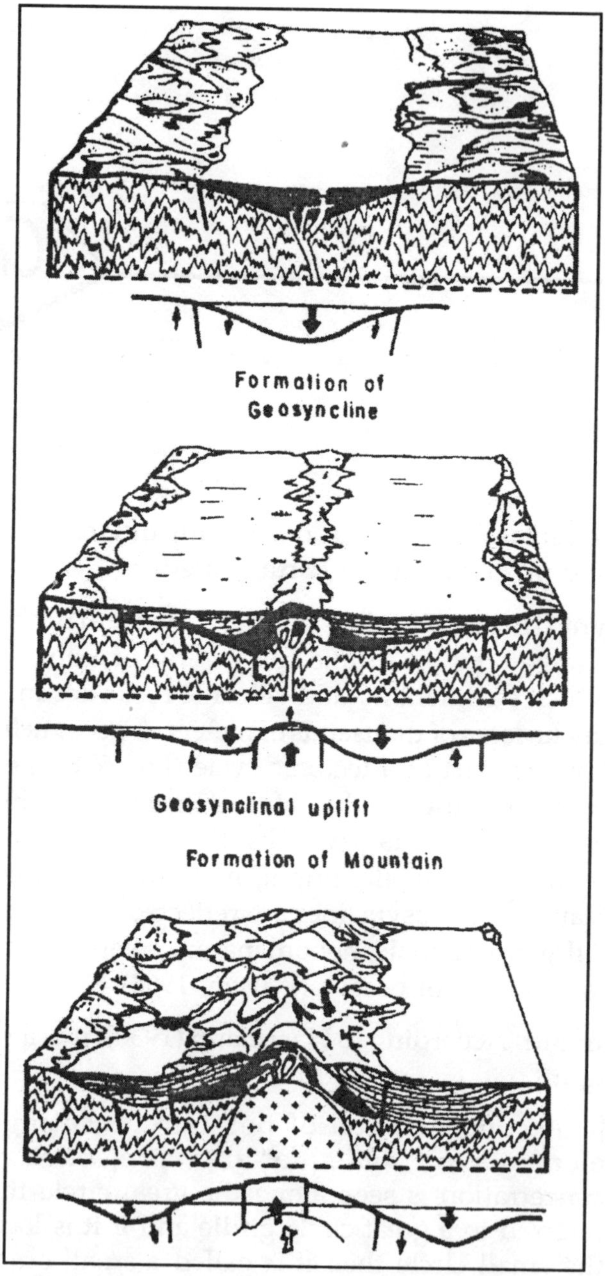

Fig. 67. Formation and uplift of geosyncline.

Gliebretter *(shear fold). In this type to folding the rock appears to be sliced.* The rock is traversed by narrow and spaced shear planes or bands affected by deformation (Fig. 68).

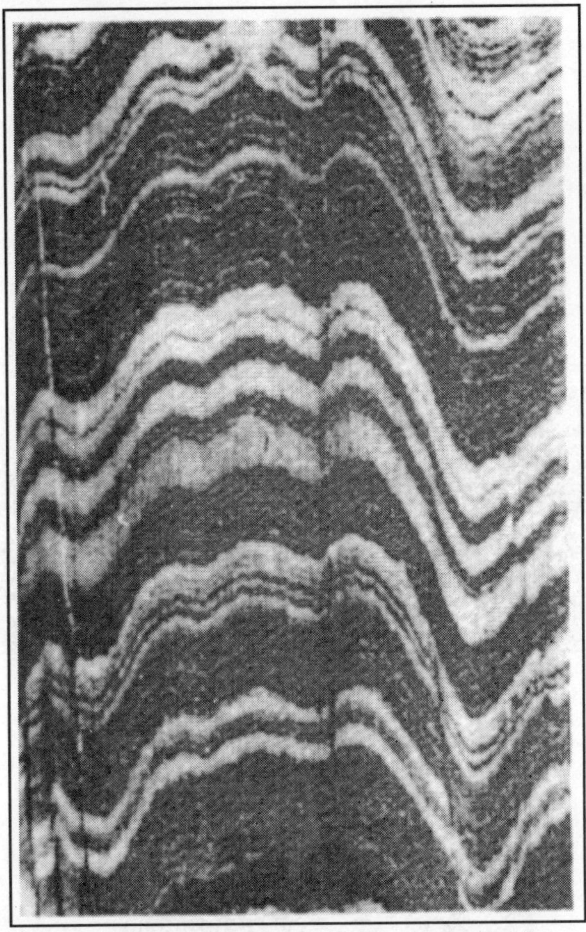

Fig. 68. Shear folding in slate from 'Centre country' (crest of fold), Bendigo. Natural size, in Hills, 1973.

Glide plane : glide reflection; translation plane; gliding plane; slip plane [cryst]. A symmetry plane in a crystal characterized by translation parallel to the plane.

Glide twin *(deformation twin).*

Graben : trough fault A depressed crustal unit or block bounded by faults on its long sides. (Fig. 69 ABC).

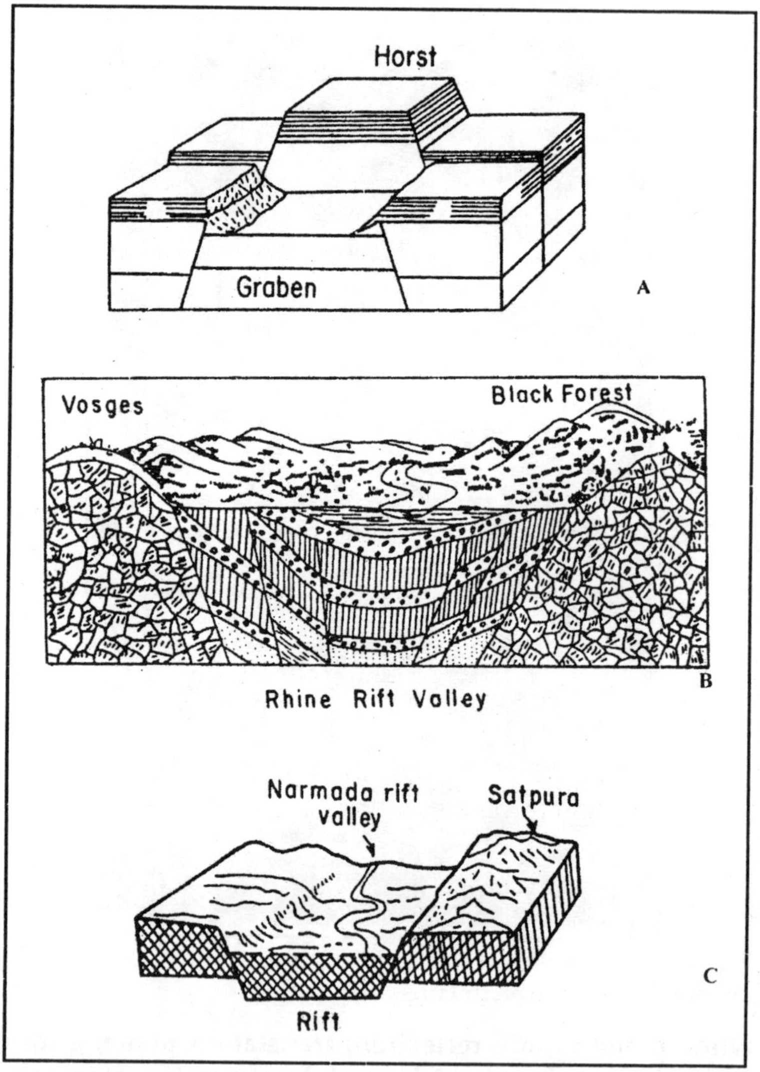

Fig. 69. Horst and Graben structures (after Lahee, 1971) (b) Rhine Rift Valley (after Lange et al., 1966) (c) Narmada Rift Valley (Valdiya, 1984).

Gravitational gliding : *gravitational sliding.*

Gravitational sliding Due to the effect of gravity the rock masses move downward along the slope. (e.g. thrust plane). (Fig. 70).

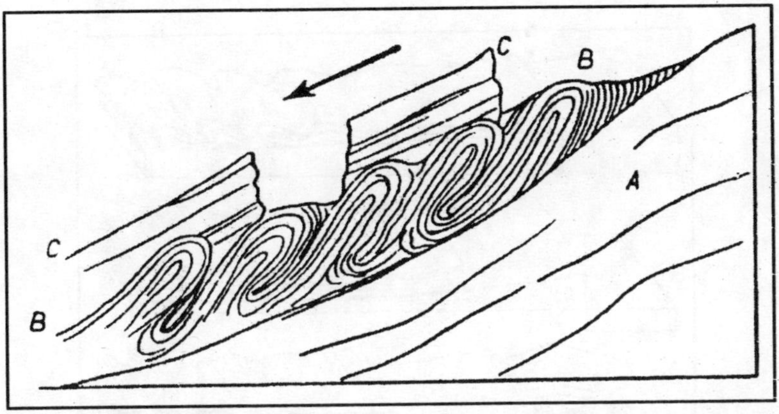

Fig. 70. Gravitational Tectonics, according to Naumann (in Hills, 1973).

Gravity fault : *normal fault.* (Fig. 71). In this type of folding the hangingwall block goes down relative to the footwall-block.

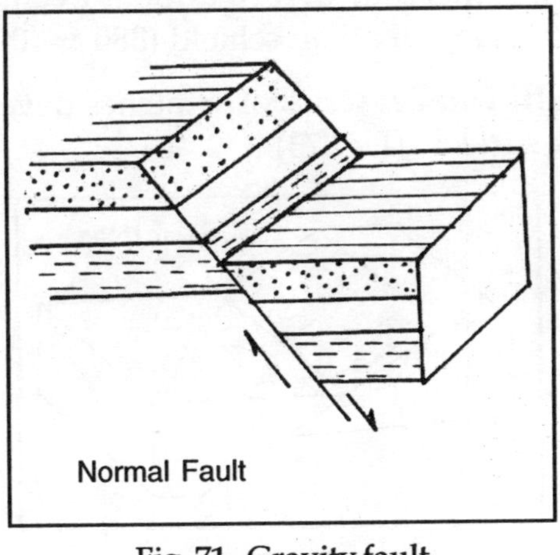

Fig. 71. Gravity fault

Gravity tectonics : Tectonics in which down-slope gliding occurs under the influence of gravity. Probably gravity movements are trigered by deep-seated crustal forces, and structures are modififed by gravity. (Reyer, in Hills, 1973), (Fig. 72).

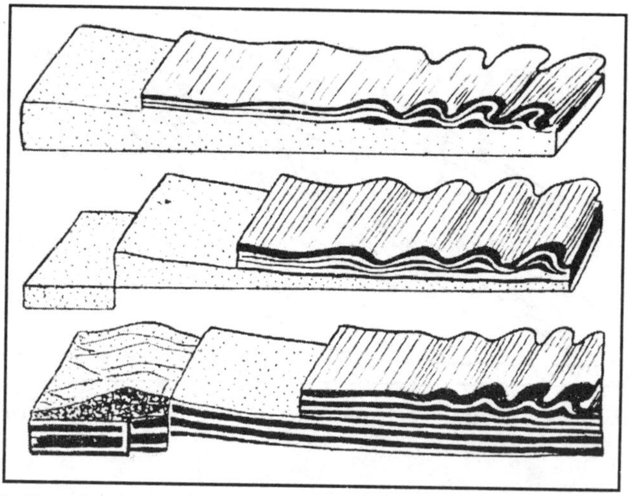

Fig. 72. Gravitational tectonics according to Reyer (in Hills, 1973).

Genville orogeny A The term is used for a major plutonic, metamorphic, and deformational event during the Precambrian Canadian schield (880 & 1000 m.y. ago).

Groove [fault] Parallel series of scratches developed along a fault surface. (Fig. 73)

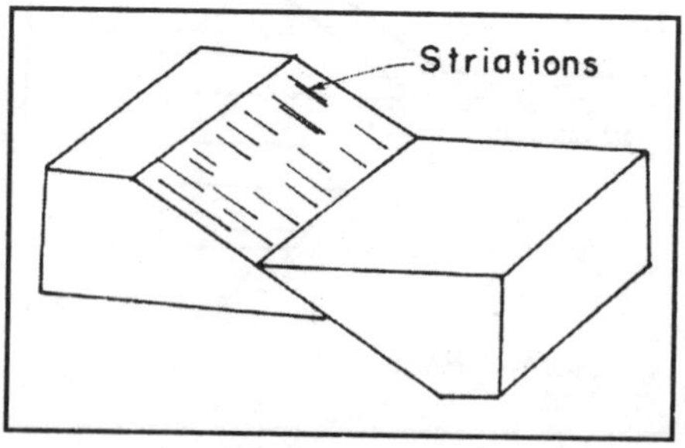

Fig. 73.

Growth fault : contemporaneous fault. Less-preferred syn: depositional fault; Gulf Coast -type fault; progressive fault; slump fault; synsedimentary fault It is a fault in sedimentary rocks associated with continuous deposition, so that the throw increases with depth and the strata of the downthrown side are thicker than the strata of the upthrown side.

Gutenberg discontinuity : Oldham-Gutenberg discontinuity; Wiecher-Gutenberg discontinuity. The seismic-discontinuity at 2900 km. represents the mantle-core boundary, at which the velocities of *P* waves are reduced and *S* waves disappear. (Fig. 74).

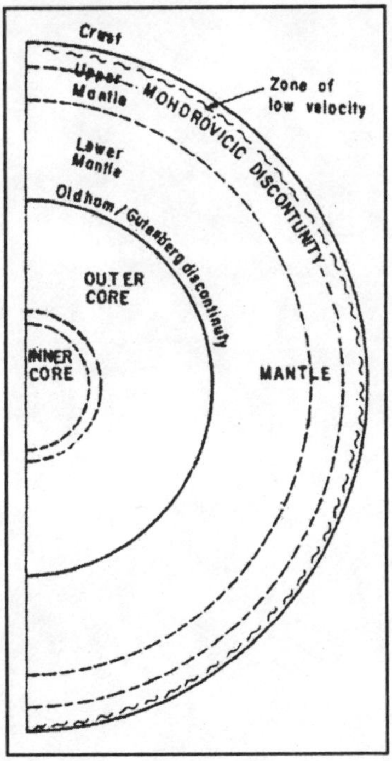

Fig. 74. Gutenberg discontinuity (after Holmes, 1965)

□□□

H

Hanging wall Ramps and Flats. During the thrusting of rocks the displacement occurs in the hangingwall and footwall like a staircase. Vertical movement of the displaced rocks is ramp and horizontal one is flat. (Fig. 75).

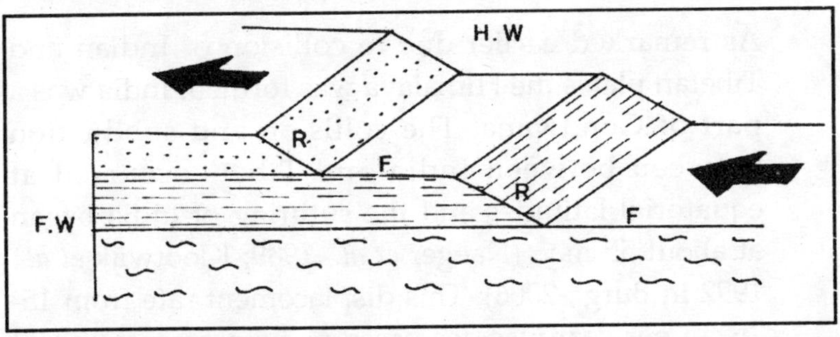

Fig. 75. (HW-Hangingwall, F.W.-footwall, R-ramp, F-flat, Arrow:transport direction).

Harpolith It is a *phacolith* with a vertical axis and is sickle shaped igneous intrusion.

Heading wall *(footwall)*.

Heave : Horizontal throw The horizontal component of separation or displacement in a fault. (Fig. 56)

Hercynian orogency : Variscan orogeny The late Palaeozoic orogenic era of Europe of Carboniferous and Permian time.

Himalayan Orogeny The world's highest mountain peak, the Everest, is situated in the Himalaya. It also has the honour of owning several mountain summits which excel in height located elsewhere on the globe. However, it is indicative of the fact that the vertically directed forces played a significant role in the later phases of the Himalayan configuration which had occurred about one million years ago.

It may be remarked that the formation of the Himalaya did not take place all of a sudden, rather its development was gradual. Its formation was initiated about 310 million years ago and this age is known as Permo-carboniferous belonging to Palaeozoic era.

As remarked earlier due to collision of Indian and Tibetan plates the Himalaya was formed. India was a part of Gondwana. The collision and subduction between between India and Tibet occurred at equatorial latitudes and the suturing of India began at about 55 m.y. (Naeger *et al.*, 1989; Klootwijk *et al.*, 1992 in Burg., 2006). This displacement rate from 15-25 cm per year was decreased to 4-5 cm per year and the shortening rate is presently 1-2 cm per year (Demets *et al.*, 1990 in Burg, 2006). Patriat and Achache (1984) described about the movement of India and Eurasia which occurred between 61 and 59 m.y. (Fig. 76).

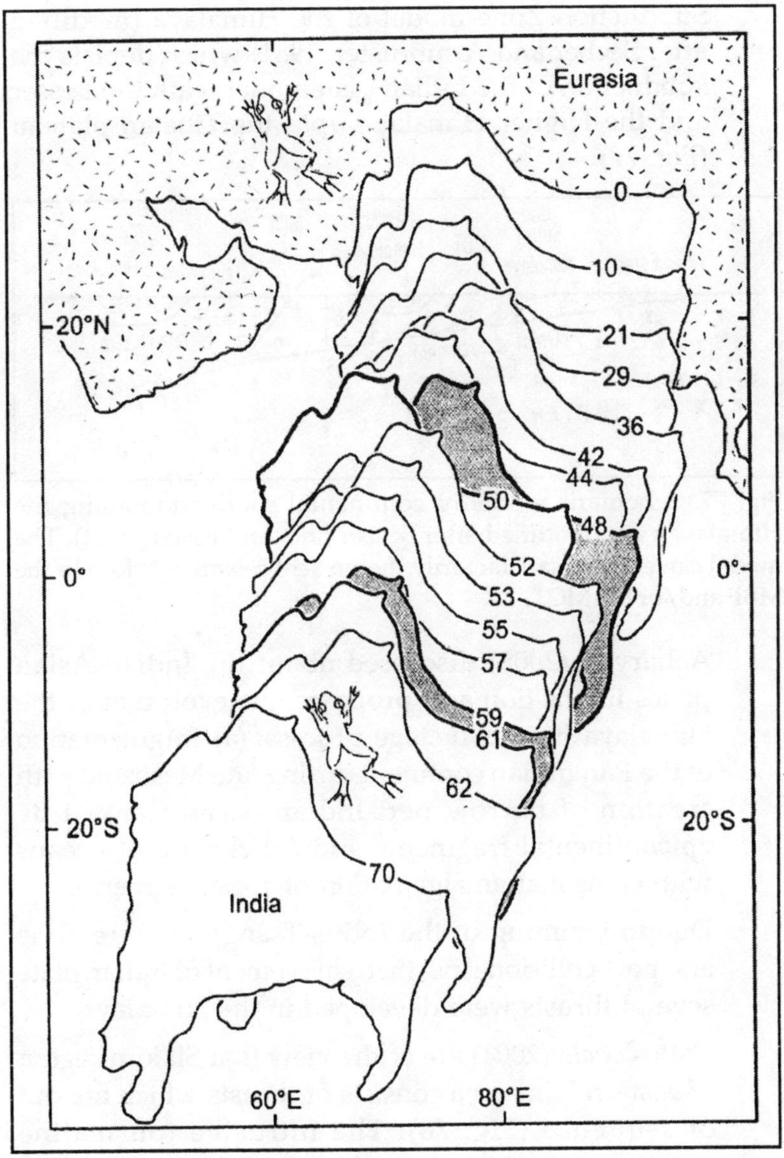

Fig. 76. Northward drift of India with respect to Eurasia since 70 Ma (after Patriat and Achache, 1984). Numbers on the northern boundary of India indicate time in Ma for that position. Frogs refer to 65 Ma old continental bridging between India and Asia.

Subduction Zone model of the Himalaya (modified after Seeber and Armbruster, 1981) which depicts the subduction of the Indian plate underneath the Lesser and the Higher Himalaya upto the Tibetan plateau (Fig. 77).

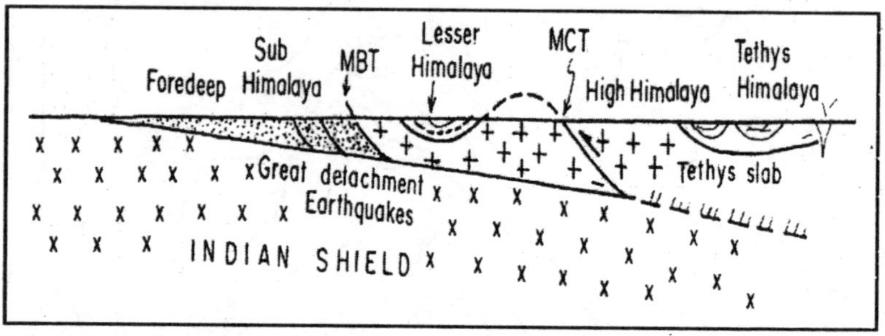

Fig. 77. Schematic model of continental subduction along the Himalayan arc (modified after Seeber and Ambruster, 1981). The model can explain satisfactorily the present seismicity close to the MBF and/or the MCT.

Acharyya (2005) discussed about the Indian-Asian plates interaction and proposed the evolution of the Himalaya by a two stage process; (a) fragmentation of the Pan Indian continent during late Mesozoic with creation of narrow peri-Indian ocean flanked by epicontinental fragments and (b) closure of oceans with collisonal amalgamation of rock-fragments.

Due to jamming of the Indus-Tsangpo suture zone and post-collisional northern movement of Indian plate several thrusts were developed in the Himalaya.

Catlos *et al.*, (2004) are of the view that Sikkim region of eastern Himalaya consists of thrusts which are out of sequence (Fig. 78). The intrusive tourmaline granites of the Himalaya (14.8 ± 0.3 Ma and 2.0 Ma) is a melt formation within the Himalayan orogenic belts. Leucogranites are exposed near Gangotri, Manaslu, Sikkim *etc.*

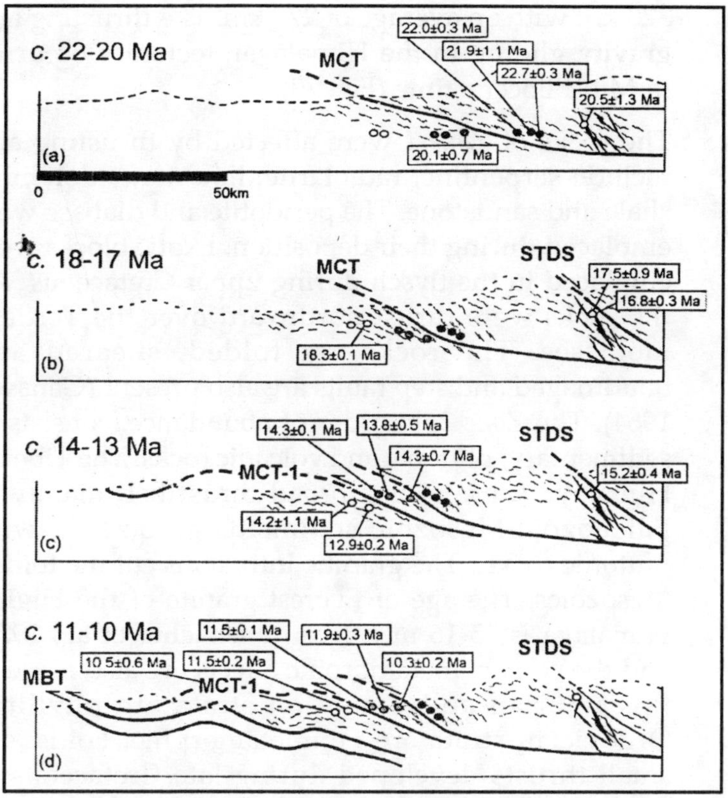

Fig. 78. Schematic interpretation of the age results and P-T constraints from Sikkim rocks. Greater Himalayan Crystalline-rocks (samples NLG 963, LCG 542 and LCG 541) are indicated by black circles, and the Lesser Himalaya samples (KBP 1062A and KBP 1062 C) are white. The High Himalayan leucogranite MK 51K is indicated in white as well. (After Catlos *et al.*, 2004).

Saklani (1993) established that the MCT zone is characterized by imbrications or schuppen structures and the varying ages of the crystallization are due to imbrications which pile over one another forming Duplex structures.

Himalayan Tectonics The Himalayan orogen measures about 2400 km in length with a width of about 230 to

320 km with an average of 270 km. The thrusting and gravity gliding in the Himalayan tectonics occurred in Mio-Pliocene time (Fig. 79).

The rocks of *Zone-1* were affected by thrusting and include serpentine, radiolarite, limestone, dolomite, shale and sandstone. The peridotite and diabase were emplaced during their deposition. Exotic blocks were emplaced in the flysch during upper Cretaceous. All the rocks were thrust northward over the Tertiary molasses. The rocks are folded, sheared and boudinaged and step faults are also present (Gansser, 1964). The *Zone* 2, in order of abundance, consists of sedimentary, plutonic and volcanic rocks. The Tibetan Himalaya (*Zone* 3) is a broad thrust belt mainly of Palaeozoic-Mesozoic sediments associated with plutonic rocks. The granitic intrusions cut the folded Mesozoics. The age of Everest granite of the Higher Himalaya is 15-16 m.y. (Krummenacher et al., 1978) and the contact metamorphic effects are also present. Precambrian and Cambrian strata are phyllitic. Ordovician strata are nonmetamorphic. Folds and small thrusts developed during late Cretaceous to Pliocene. The Higher Himalaya (*Zone* 4) is a vertically uplifted region where the pre-existing sediments are eroded and consist of pre-Ordovician metamorphics intruded by granites. It consists of predominantly metamorphics with subordinate sedimentaries. The lower Palaeozoic sediments overlie the metamorphic rocks (Precambrian) and are intruded by Neogene granites. The high grade metamorphics occur at the base overlying the less metamorphosed to unmemorphosed lower Palaeozoics. Thursting movement affected the rocks associated with folds (isoclinal).

Further south of it, is the Lower or Lesser Himalaya (*Zone* 5) consisting of rocks of Precmbrian to Mesozoic

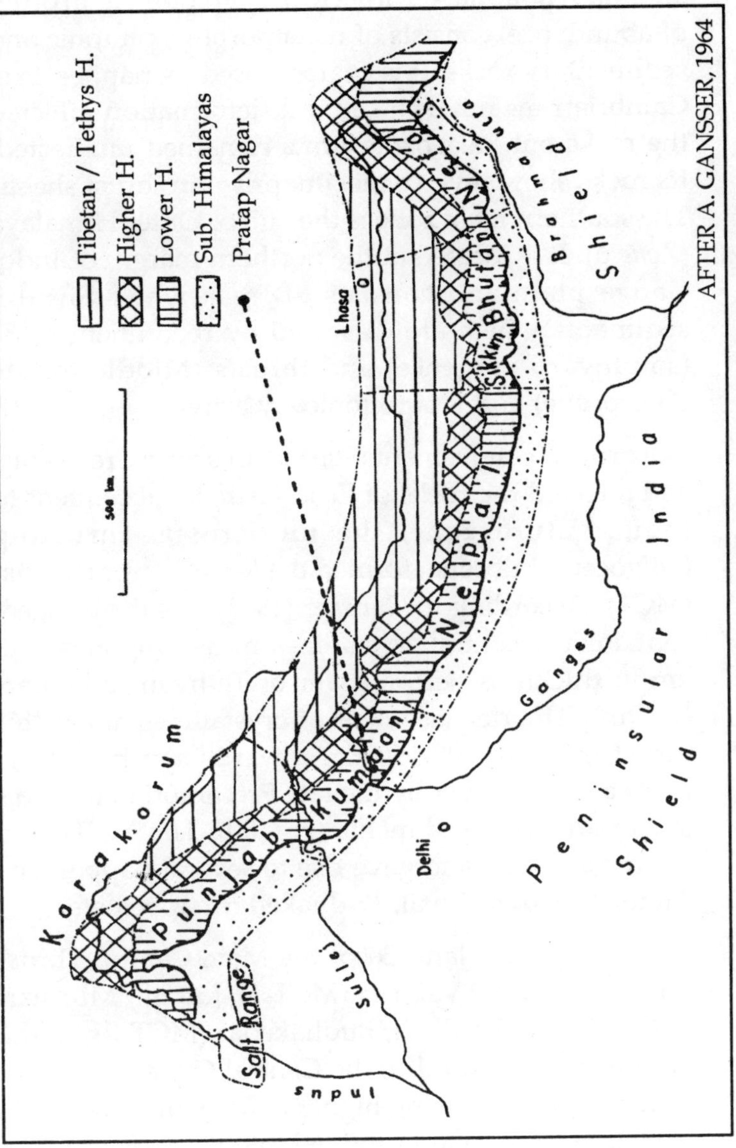

Fig. 79. Structural zones of the Himalaya (After Gansser, 1964).

age which are confined into nappes. It is sharply delinated by the Main Central Thrust (MCT) in north and Main Boundary Thrust (MBT) in south. It, in order of abundance, consists of metamorphic, plutonic and sedimentary rocks. It is characterised by nappes. Pre-Cambrian metamorphism and deformation affected the rocks but Palaeozoic strata remained unaffected. Its rocks show reverse metamoprism in thrust sheets. The southern most zone is the Sub or Outer Himalaya (*Zone 6*). It is located at the northern margin of Indo-Ganga plain and consists of fresh-water Siwalik sediments which are dissected by recumbent folds (southward vergence) and thrusts (Middle to late Pleistocene) and it is tectonically active.

The orogenic movements started in upper Cretaceous and affected the rocks of *Zone 1 and* 2. Subsequent to Indus suture zone, the intracrustal thrusting (Oligocene-Miocene) formed the Main Central Thrust (MCT). According to Gansser (1993) this thrust sheet contains the central crystallines measuring about 15 km in thickness with a cover of Tethyan sediments (12 km). The displacement of crystallines along this thrust is about 100-150 km towards south and the rocks are piled over by the late Precambrian and less metamorphosed sediments (Fig. 80). The MCT is an imbricated zone and gave rise to several nappes e.g., Almora, Jutogh, Chail, Budhakedar nappes etc.

According to Saklani (2005) the Main Central Thrust is represented by Vaikrita (MCT-1), Jutogh/Munsiari (MCT-II) and the Chail/Budhakedar (MCT III) in the form of a major duplex.The Central Crystallines and metamorphics showing high, medium and low grade of metamorphism are confined into the nappes along these thrusts. The rocks were remetamorphosed from

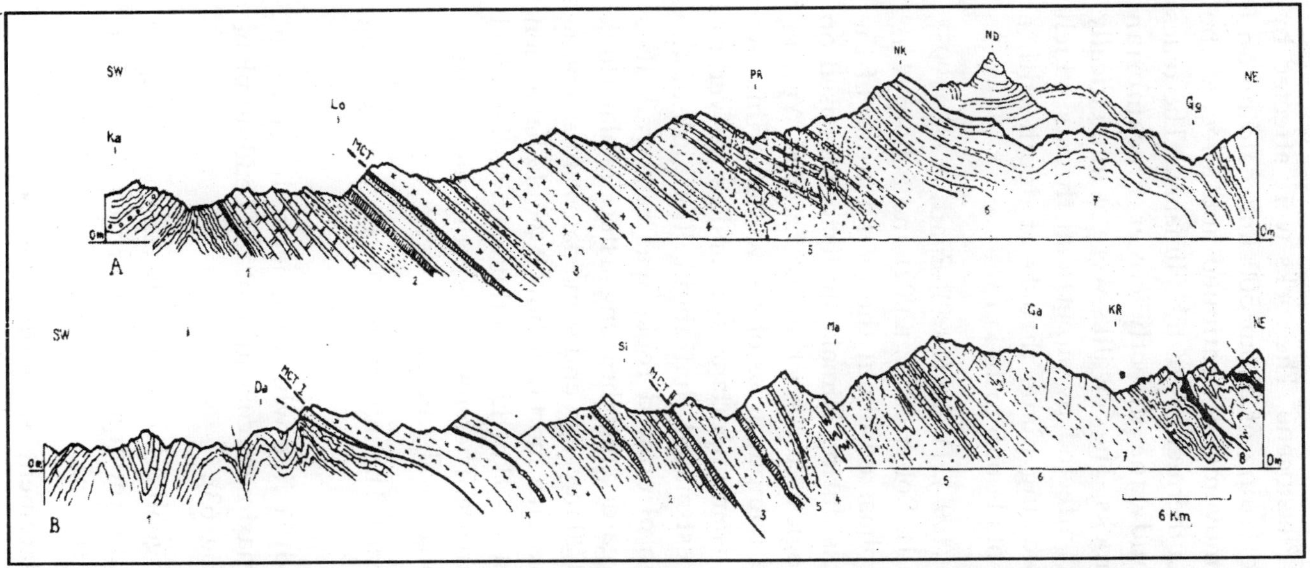

Fig. 80. *Geological sections through the MCT and the Crystalline Core of the Kumaon Himalayas* (Central Himalayas). (After Gansser 1964, 1993). Section A through Nanda Devi. Section B along Kali River (W border of Nepal), to Garbyang, 1. Precambrian sediments of Lesser Himalaya. Metamorphism increasing towards MCT. 2. Pre-cambrian metasediments of Lesser Himalayas with schists, quartzites and amphibolites just below thrust, repeated in Sirdang area below MCT II.3. Gneisses, partly migmatitic of lower part of crystalline core. 4. Metaquartzites and high grade schists in middle section of crystalline. 5. Leucogranites with dykes intruding zone of marbles, limesilicates and psammite-gneisses. The relation of leucogranites with the upper high grade carbonate zone of the main crystalline is widespread.A certain three-fold division of the main crystalline core is also evident in Nepal and the eastern Himalayas. 6. Typical biotite porphyroblast-schists, ending the main metamorphism of the crystalline core (Budhi schists). 7. Late Precambrian to Cambrian argillaceous Tethys-Himalayan sediments. 8. Folded and imbricated sediments from Ordovician to Permain of the Tethys Himalays. Da = Darchula, Ga = Garbyang, Gg = Goriganga (Valley), Ka = Kapkot, KR = Kuti Rivar, Lo = Loharkhet, Ma = Malpa, ND = Nanda Devi (7,820m), NK = Nanda Kot, PR = Pindari River, Si = Sirdang. (After Gansser, 1993).

Eocene to Pleistocene. The rocks were affected by crustal shortening of about 500 km. Folding and thrusting movements continued and affected the molassic southern edges of the Himalaya. The relics of the Gondwana directly cover Precambrian basement rocks. The Siwaliks were asymmetrically folded and affected by superficial thrusts which become bedding plane thrusts at depth with a shortening of about 5 to 10 km.

Steck *et al.* (1993), Dezes (1999), Epard *et al.*, (1995), Girard (2001), Robyr et al., (2002) and Schlup *et al.*, (2003), emphasized about the over burden of the eroded rocks during Himalayan-tectonics based on many geological cross-sections of Zanskar, Yunam, Manali and Rampur areas of the western Himalaya (Fig. 81). It is an important issue and as per vertical scale it is estimated that about 30-40 km of the overburden of eroded rocks was removed. It is generally regarded that thrusting in the Himalaya took place about 20 m.y. ago and some dates are as young as 5 m.y. In such a short time how it would have been possible that 30-40 km thick overburden was deposited in the Indian sea. Therefore these estimates appear to be more conjectural and exaggerated.

Hinge [fold] flexure The maximum curvature or bending in a fold.

Hinge line [fold] A line which connects the points of flexure or maximum curvature of the bedding plane in a fold. (Fig. 63).

Hinge line [struc geol] A boundary between a stable region and a region undergoing upward or downward movement. (Fig. 63)

Horizontal displacement *(strike slip).* (Fig. 56)

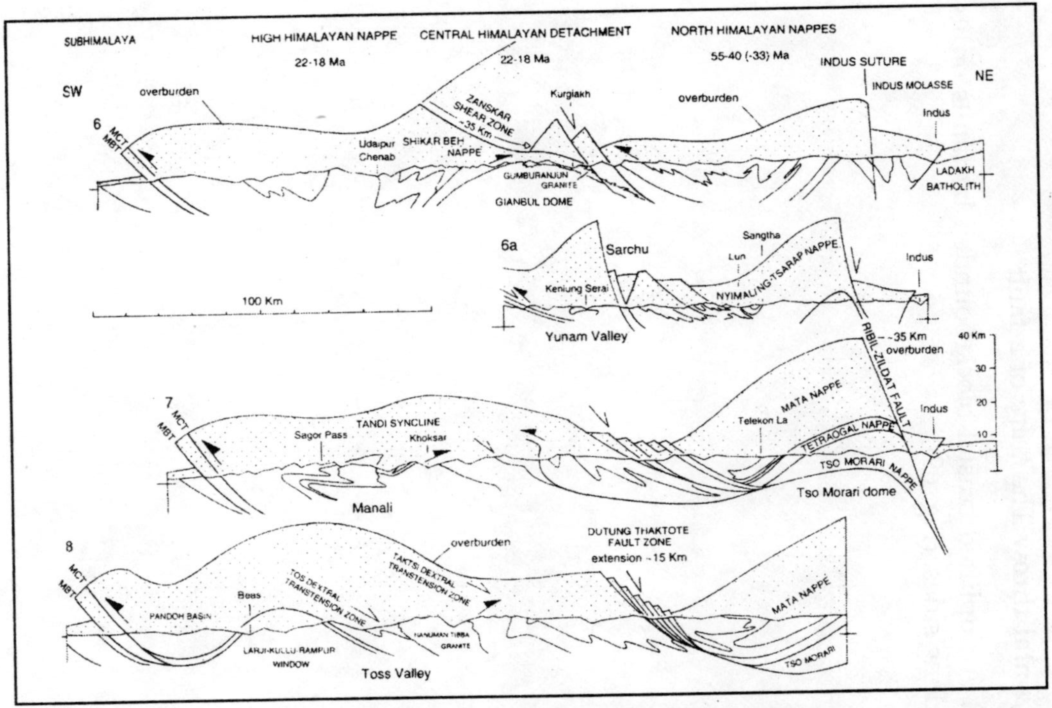

Fig. 81. Estimates of the eroded overburden of the Himalayan accretionary wedge based on thermo-barometric data from Dezes (1999), Epard et al. (1995). Girard (2001), Robyr et al. (2002), Schlup et al. (2003), and Steck et al. (1993). The numbers 6, 7and 8 correspond to the cross-sections in the Yunam Valley and the Tsara River near Lun and Sangtha. Note the strain partitioning of the Central Himalayan detachment along the strike of the Himalayan range.

Horizontal fold *(nonplunging fold).*

Horizontal slip : horizontal dip slip A horizontal component of the net slip in a fault.

Horizontal throw The *heave* of a fault.

Horst An uplifted crustal block bounded by faults on its long sides. (Fig. 69).

□□□

I

Idiogeosyncline A type of geosyncline between stable and mobile areas of the crust, the sediments are folded, (Umbgrove, 1933).

Igneous breccia A breccia which is composed of fragments of igneous rock due to Igneous processes (e.g. volcanic breccia, intrusion breccia).

Imbricate structure : schuppen structure; shingle block structure This structure is characterized by a series of nearly parallel and overlapping thrust faults, high angle reverse faults, or slides, associated with rock slices, sheets, plates, blocks, or tectonic horses which have nearly the same displacement (Fig. 82).

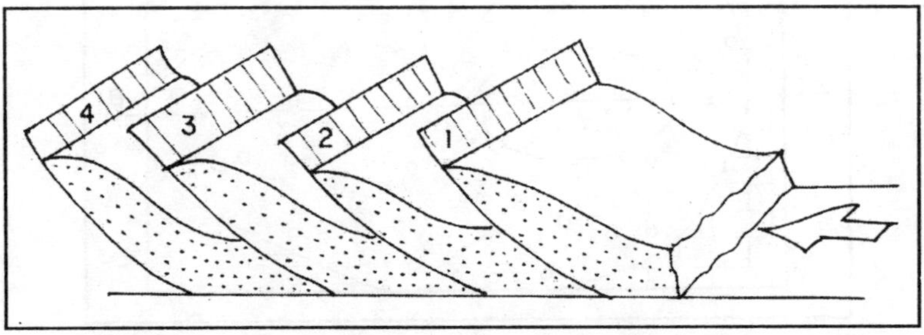

Fig. 82. Tectonic horses: 1 to 4 (Imbricate Structures).

Imbrication [tect] The steeply inclined, overlapping arrangement of thrust sheets.

Inclined fold When an axial surface of a fold is inclined from the vertical, and one limb is steeper than the other.

Incompetent Rocks deformed in ductile conditions as compared to adjacent more brittle rocks. *Similar folds and Parallel folds are commonly developed in such rocks.*

Incremental strain The total finite strain of a rock body is the cumulative result of a number of incremental strains.

Inlier The order rocks surrounded by the rocks of younger age.(= klippe). (Fig. 83, 84).

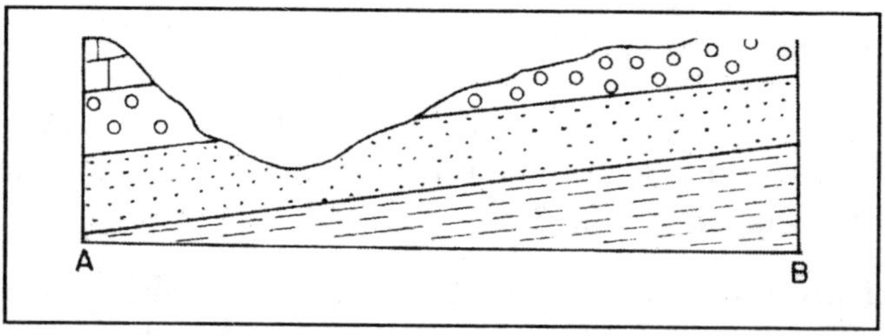

Fig. 83.

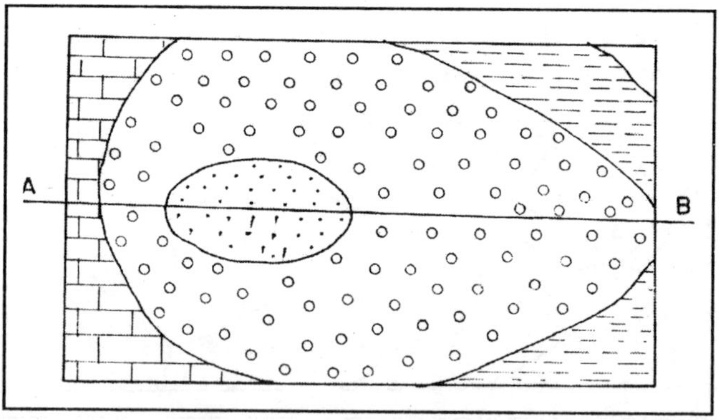

Fig. 84.

Interfolding The simultaneous development of discrete fold-systerms with different orientations.

Intrafolial fold Folds developed only in a few layers otherwise unfolded rocks. (Fig. 85).

Fig. 85.

Intraformational (a) Formed within a geological formation, more or less contemporaneously with the enclosing sediments. The term is used with regard to syndepositional folding or slumping, e.g. "intraformational deformation" or "intraformational breccia". (b) Existing within a formation, with no necessary cannotation of time of orgin.

Involution [struc geol] Refolded thrust sheets/nappes, resulting in a complex pattern.

Irrotational strain : nonrotational strain In this type, the orientation of the principal axes of strain remains unchanged.

Isopach : ispachyte; thickness line; thickness contour. A line sharing points of equal true thickness of a designated stratigraphic units.

Isopach map : thickness map A map that shows thickness of a bed, throughout a region with the help of isopachs at regular intervals.

Isostascy : The crustal continents are lighter than the mantle and float over the latter. Earlier it was believed that high and small mountains were blocks of different densities floating above the same base line (Fig. 86.1). However, it is now thought that the rock-blocks of same density but of different heights and thickness float at different depths (Fig. 86.2).

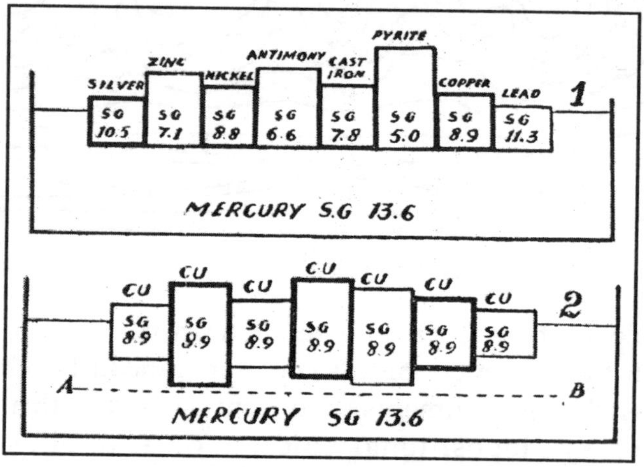

Fig. 86. Illustrations of isostasy

Isostatic anomaly A type of anomaly in which the gravitational effect of masses extending above sea level is approximately compensated by a deficiency of density of the material beneath the masses.

J

Joint A plane surface of fracture or parting in a rock, without displacement.

Jointing The presence of *joints* in a rock which can be tensional, longiudinal and oblique (Fig. 87).

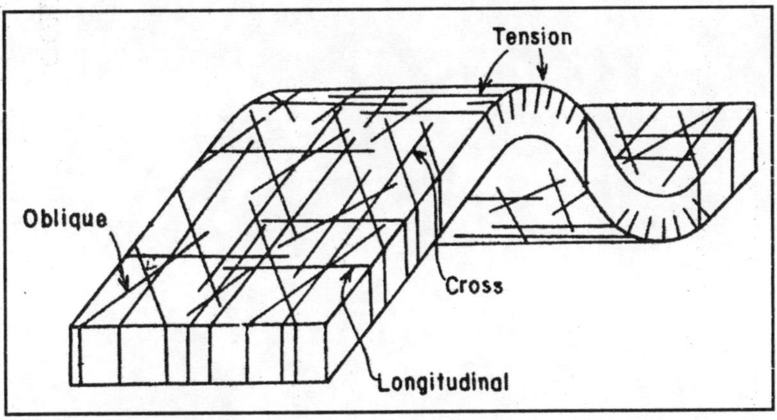

Fig. 87. Various types of joints (after Whitten and Brooks, 1972).

Joint plane The surface of a joint.

Jura-type fold *(décollement fold).*

□□□

K

Keystone fault A graben type structure developed on the crestal part of an anticline.

Killarney Revolution A supposed major orogeny which took place at the end of Precambrian time in North America.

Kimmerian orogeny One of 30 or more short-lived orogenies during Phanerozoic time postulated by Stille (1930-1936).

Kink fold A fold sharing sharp angular hinge. (Fig. 88).

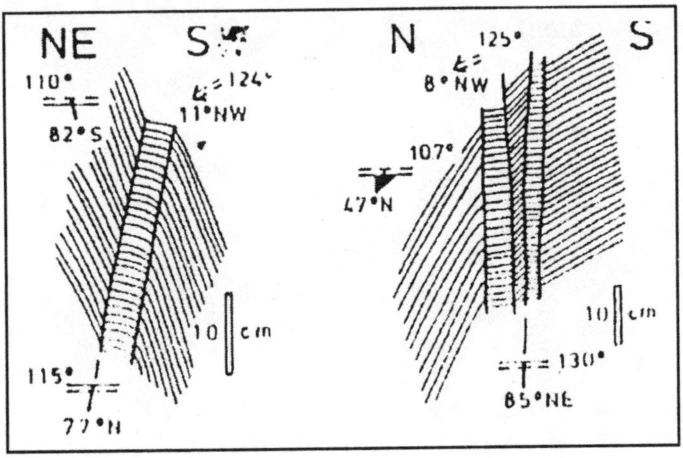

Fig. 88. Kink band structures (F_4) in slate-quartzitic beds of Infra-Krol; 6 and 7 km south of Narendranagar, Garhwal Himalaya (After Schwan, 1980).

Klippe An erosional remnant of a *nappe thrust sheet.* Auden (1937) described the windows and klippen structures from Tehri-Garhwal Himalaya (Fig. 89).

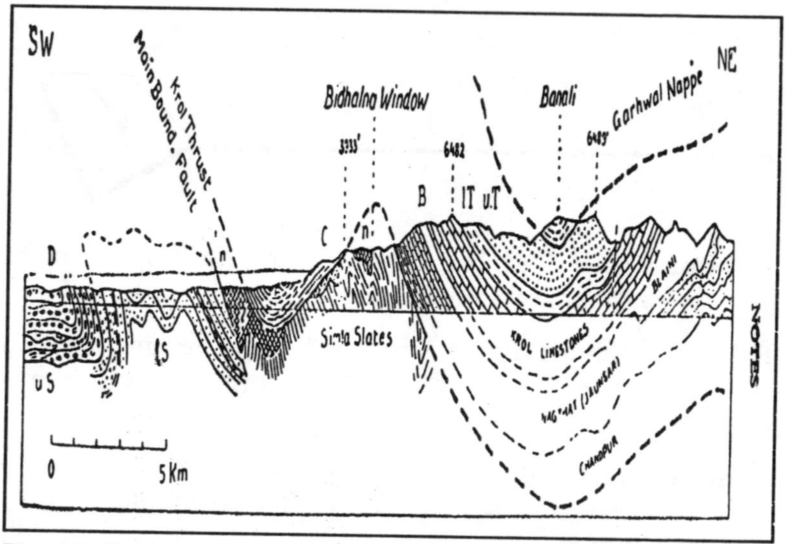

Fig. 89. Geological section through the Garhwal Himalaya after Auden (1937) showing the Garhwal thrust bound Banali Klippe

Kraton *(craton).*

NOTES

L

Lag fault : lag deposit In an overthrust, the thrusted rocks which move differentially and the upper part of the geologic section is left behind then the upper limb of an overturned anticline is represented by a *lag fault* (Fig. 90).

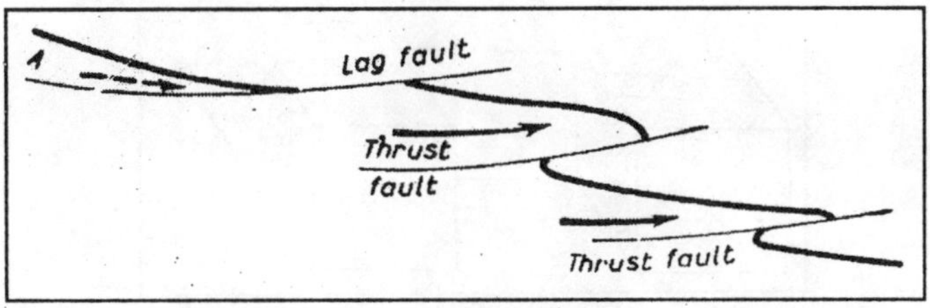

Fig. 90. Lag fault

Laramide orogeny : laramic orogeny; Laramian orogeny. Orogeny of the eastern Rocky Mountains of the United States, and its several phases extending from late Cretaceous to the end of the Palaeocene. Intrusives and mineral deposits emplaced are commonly called as Laramide (e.g. the Boulder batholith, Montana). Opinions differ to use term Laramide to a single event or to apply it to all orogenies from early in the Cretaceous through the Eocene or later, of the whole Cordilleran belt of western North America. The Laramide is an orogenic era in the sense of Stille (1930-1936). The Laramide Formation of Wyoming and Colorado is probably a synorogenic deposit.

Lateral fault A fault showing horizontal strike separation. Also the *stirke-separation fault* (Fig. 91).

Lattice-preferred orientation The preferred orientation of crystallographic axes or planes. due to *plastic deformation* and recrystallization It is controlled by the mineral structure and pressure, temperature, stress during deformation.

Left - lateral fault : sinistral fault; left-lateral silp fault; left-silp fault. A fault on which the displacement is *left-lateral.* (Fig.91).

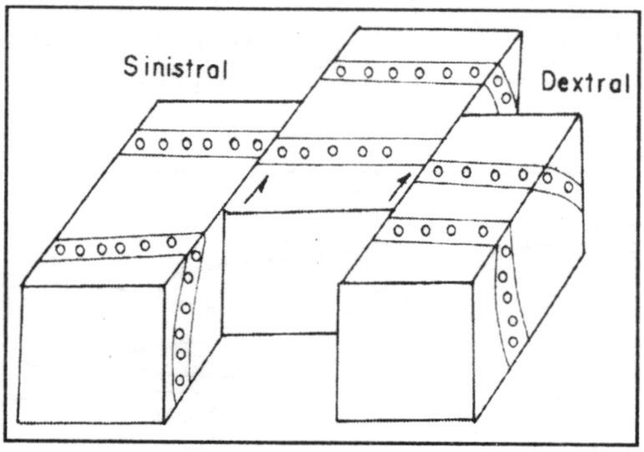

Fig. 91. Lateral fault

Level fold *(nonpluning fold).*

Left joint A tension joint in massive rock formed as a result of reduction of load pressure during quarrying. It is a type of *strain break.*

Limb [fold] *(flank).*

Lineament A regional linear topographic feature reflecting a crustal structure (Hobbs *et al.,* 1976). e.g., as fault lines, and straight river courses. (Bajpai and Kandwal, 2005; Fig. 92).

Fig. 92. (A) Lineaments and their relation with channel pattern of the Yamuna River (Y) in Delhi (D), Faridabad (FD), Region. S-Sihi, B-Ballabhgarh, H-Hindan River, Ri-Ridge Lineament, Ff-fault, Qtz-quartzite. (After Bajpai and Kandwal, 2005).

Linear element The Lineations are the common linear elements. (Fig. 93).

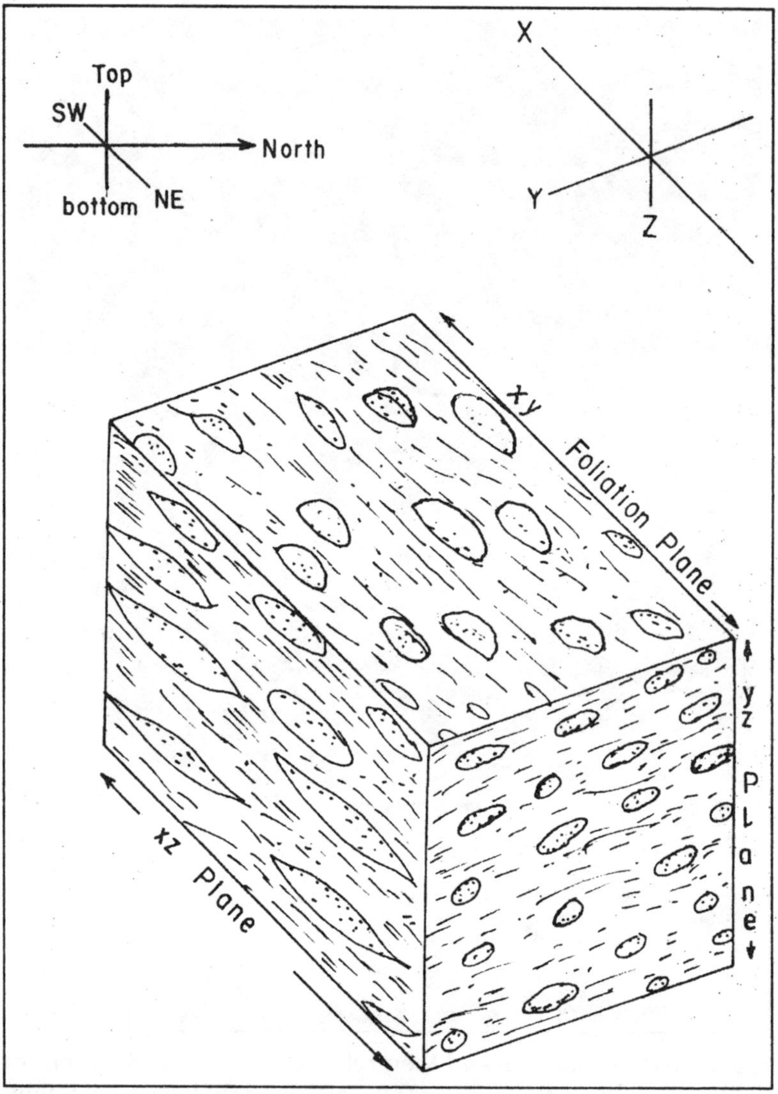

Fig. 93. Schematic block diagram showing linear structures formed by mineral-grains on the principal plane.

Lineation A nongenetic term for any linear structure in a rock such as slickensides and axes of folds etc. Lineation in metamorphic rocks includes mineral streaking and crinkles and minute folds parallel to fold axes, and the lines of intersection between bedding and cleavage (Cloos, 1946).

Lithosphere In plate tectonics, it is a layer underlying the *asthenosphere.* It includes the *crust* and part of the *upper mantle* and measures for about 100 km in thickness (Dennis & Atwater. 1974).

Load fold A plication of an underlying strata which results from unequal pressure and settling of overlying material.

Local unconformity An unconformity limited to geographic extent representing a relatively short period, It lacks the regional importance.

Longitudinal fault A fault whose strike is parallel to the structural trend of the region (Fig. 94).

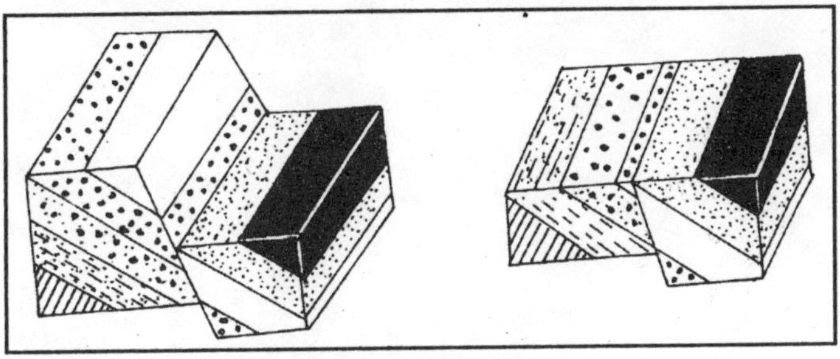

Fig. 94. Longitudinal fault

Longitudinal fold : strike fold A fold whose axis trends with the general strike of the structural set up of an area.

Longitudinal joint : S-joint, bc-joint. A steeply dipping joint plane in a pluton that is oriented parallel to the flow lines.

Low : structural low A general term for a structural basin; a syncline, a saddle, or a sag.

Low-angle fault A fault dipping less than 45°.

Lower plate The footwall of a fault.

L-tectonite A tectonite containing lineations, (e.g. a deformed conglomerate showing elongateed pebbles). (Fig. 95).

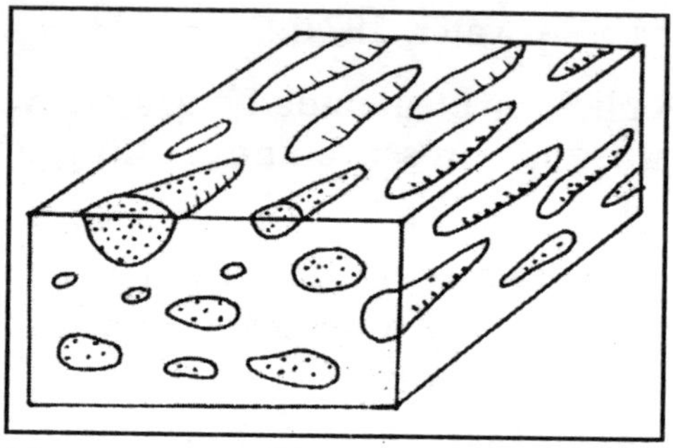

Fig. 95. L-tectonite consisting of elongateed pebbles.

□□□

M

Macro-axis The longer lateral axis (b axis)of an orthorhombic and/or triclinic mineral.

Macrofabric *(megafabric).*

Macroscopic (*megascopic)* According to Dennis (1967) The tectonic features which are large to be observed directly are macroscopic.

Magnetic dip *(inclination: magnet).*

Major fold A large-scale fold associated with *minor folds.*

Major joint *(master joint).* A joint greater-than of average extent.

Megabreccia It is a rock produced by large scale brecciation, containing randomly oriented blocks which can be inclined and mesured for more than 100 m in horizontal dimension. Longwell (1951) used it for coarse breccia containing shattered blocks 400m long, developed downslope of thrusts /gravitational sliding. These are tectonic or sedimentary in origin.

Megatectonics : geotectonics and global tectonics The tectonics of large scale structural features of the Earth.

Melange A rock characterized by the fragments and blocks of all sizes, exotic and of native origin and are embeded in a fragmented and generally sheared matrix. It can be either an *olistostrome* (sedimentary mélange) or a tectonic mélange". Gansser (1964) described its occurrence in rocks of Ladakh area of the Himalaya.

Mesogeosyncline : mediterranean A geosyncline between two continents receiving their clastic sediments.

Mesoscopic Dennis (1967), introduced it for a tectonic feature large enough to be observed without the aid of a microscope yet small enough which can be observed directly.

Microtectonics (of *structural petrology).*

Mid-oceanic ridge A seismically active, median mountain range present in Atlantic and Pacific occean. Mid Atlantic Ridge measures 25 to 50 km in width and is deep from 250 to 750 m. It is a rift valley with a rugged topography. The mid-oceanic ridge is the source of new crustal material. *(mid-ocean rise; oceanic ridge)* (Fig. 96).

Miogeosyncline : miomagmatic zone A geosyncline in which volcanism is not associated with sedimentation; (Stille, 1940).

Mobile belt A long, narrow crustal area of tectonic activity. The belt can be of *geosynclinal origin.*

Mohorovicic discontinuity : Moho; M-discontinuity. It is a type of seismic-velocity discontinutiy separating the Earth's crust from the mantle. Here the *P*-wave velocities change abruptly from 6.7-7.2 km/sec (in the lower crust) to 7.6-8.6 km/sec or average 8.1 km/sec (at the top of the upper mantle). The depth ranges from about 5-10 km beneath the ocean floor to about

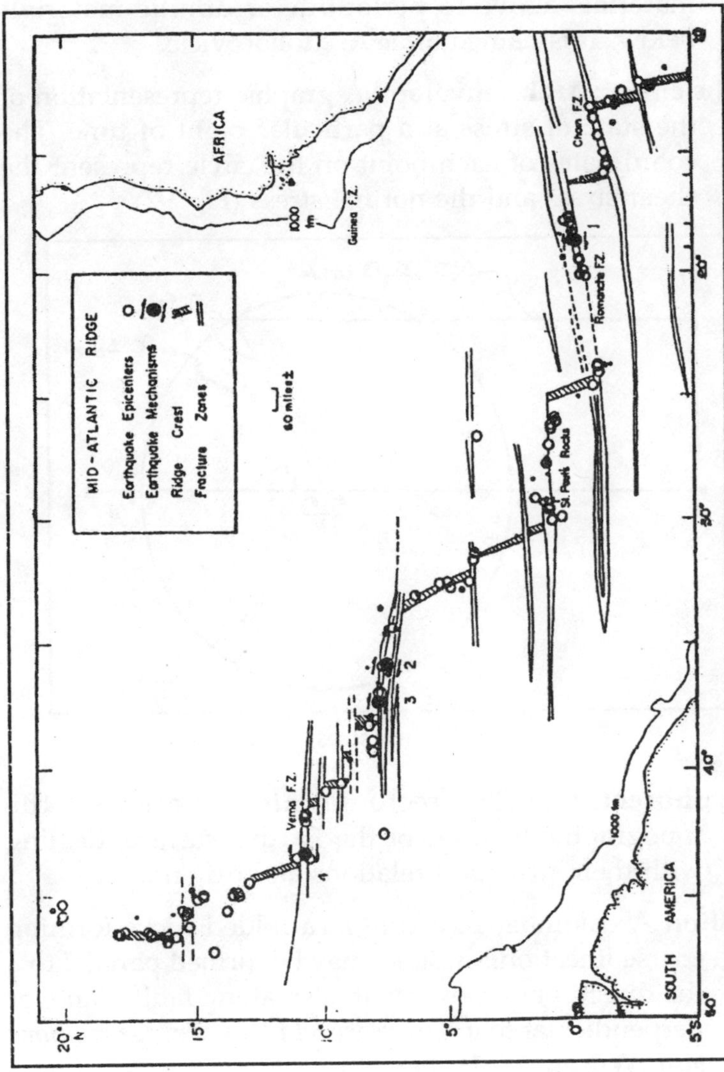

Fig.96. Transform faults on Mid-Atlantic Ridge. (After Sykes, in Billings, (3rd edition, 2000).

35 km below the continents, reaching up to 60 km under the mountain ranges. The discontinuity also represents a chemical change from basaltic or simatic materials above to peridotitic or dunitic materials below. It is named after A. Mohorovicic.

Mohr circle : Mohr envelope A graphic representation of the state of stress at a particular point of time. The coordinates of each point on the circle represent the shear stress and the normal stress (Fig. 97).

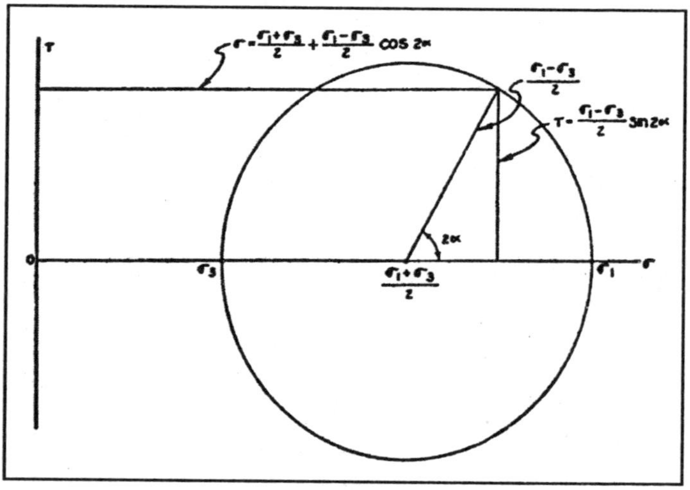

Fig. 97. Mohr's stress circle.

Morphotectonics The tectonic interpretation of the topographic features of the Earth's surface; dealing with their structural relations and origin.

Mullion A columnar structure in a folded rocks forming coarse lineation.Mullions may be formed parallel to a the direction of movement, also along fault plane, or perpendicular to it. There are *fold* and *cleavage mullions also* (Wilson, 1961) (Fig. 98).

Mullion structure A wavelike pattern of parallel grooves and ridges, formed on a folded or a faulted surface.

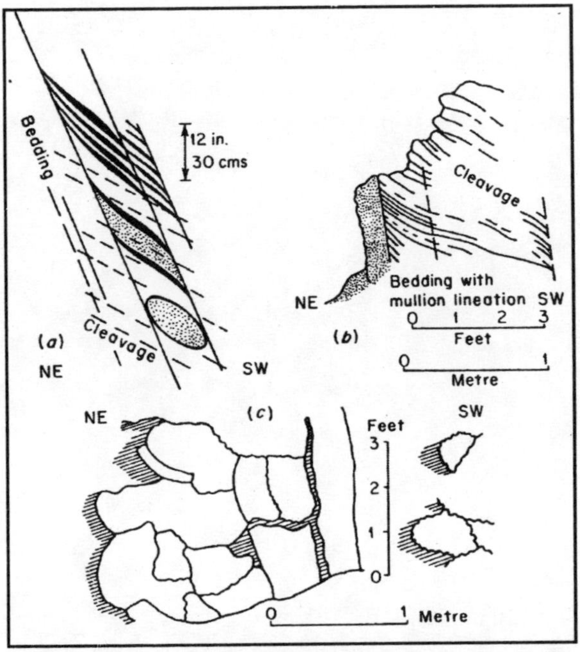

Fig. 98. (a) and (b) The development of cleavage mullions, with examples; (c) profiles of irregular mullions. (From Wilson, 1953; 1961).

Multiple fault *(step fault).*

Mylonitic structure A sturcture produced by strong brecciation and shearing which appear like flow structures. (Fig. 99).

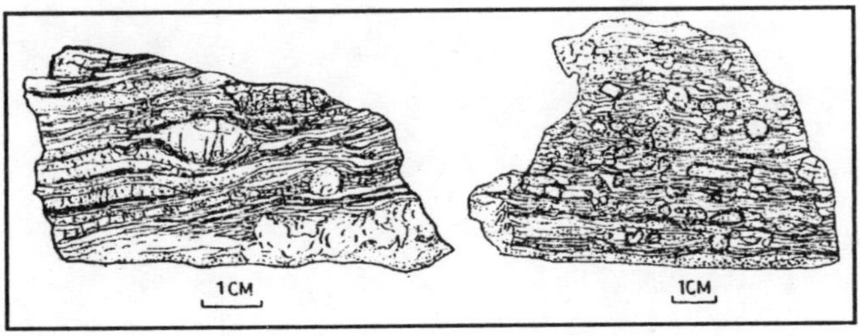

Fig. 99. Sketches of Mylonites from Ghuttu-Garhwal Himalaya (After, Saklani, 1993).

□□□

N

Nappe : Decke A sheetlike , allochthonous rock which moved on a horizontal surface. The mechanism was first described by Schardt (1893) fot the allochthonous rocks of the Western Alps (Fig. 100).

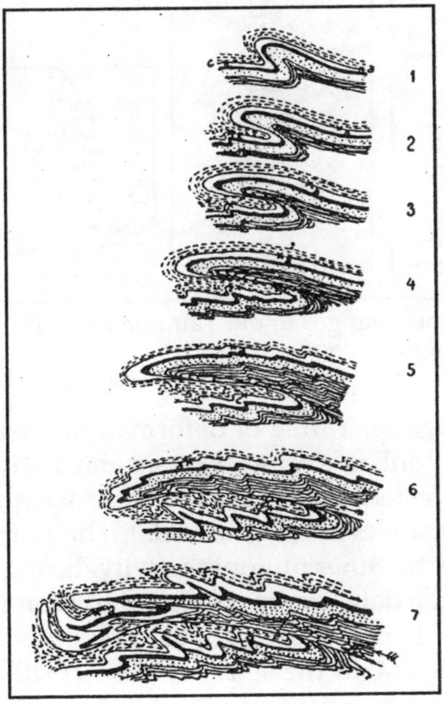

Fig. 100. (1) Overfold; (2) to (4), recumbent anticline; (5) to (7), true nappes.

Nappe outlier (klippe).

Neotectonics The study of the post-Miocene sturtures and structural history. Bajpai and Kandwal (2005) described the changes in the channel of the Yamuna river from, 1799-1999 near Faridrbad-Delhi. (Fig. 101).

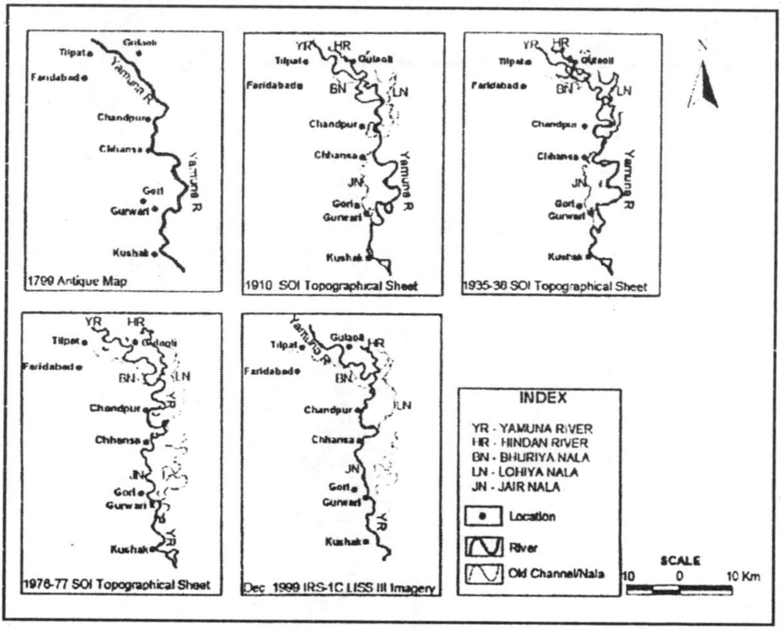

Fig. 101. Channel changes in the Yamuna River (1799-1999) After Bajpai and Kandwal, 2005.

Nevadan orogeny A time of deformation, metamorphism, and plutonism during Jurassic and Early Cretaceous in the western part of the North American Cordillera, the Sierra Nevada, (California). The emplacement of granite and other plutonic activity during the orogeny have been dated radiometrially between 180 m.y. and 80 m.y. It can most properly be considered as an orogenic era, in the sense of Stille (1940).

New global tectonics A general term for *global tectonics* which deals with the global analysis of the relative motions of crustal segments delineated by the major seismic regions.

Nonconformity : heterolithic unconformity. An *unconformity* developed between sedimentary rocks and older rocks (plutonic igneous or massive metamorphic rocks) which are subjected to erosion (Fig. 102).

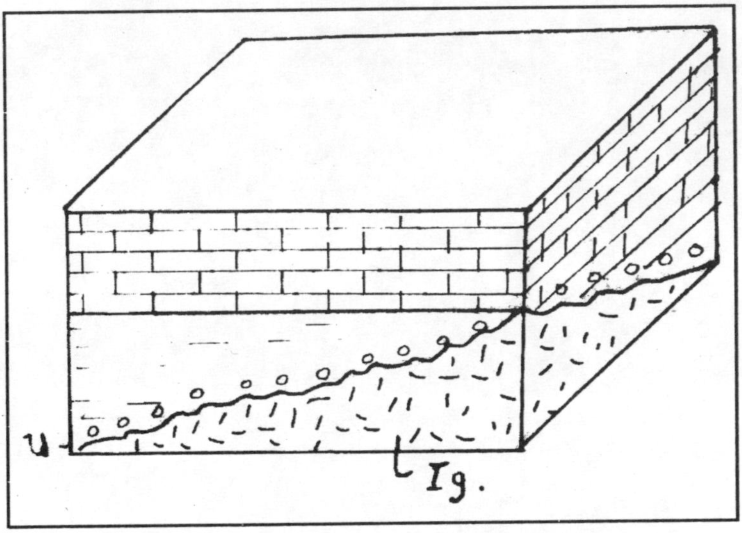

Fig. 102. Hetrolithic Uncon formity (u), Ig-Igneous)

Nonplunging fold : horizontal fold; level fold. A fold whose hinge line is horizontal.

Nontectonite A rock fabric formed by mechanical settling. Some rocks are transitional between a tectonite and a nontectonite (Turner and Weiss, 1963).

Normal dip *(regional dip).*

Normal displacement *(dip slip).*

Normal slip fault *(normal fault).*

Normal strain Change per unit length in a given direction.

Normal stress The *stress* which is perpendicular to a given plane which can be *tensile* or *compressive.*

Nose [fold] structural nose; anticlinal nose A plunging anticline without a closure.

❑❑❑

O

Obduction It is a phenomenon where oceanic crust is overthrust on to the continental lithospheric plates.

Oblate ellipsoild An ellipsiod that is flattend at the poles.

Oblique fault : diagonal fault. A fault that strikes oblique. It displaces the strikes of the rocks or dominant structure. (Fig. 103 a, b)

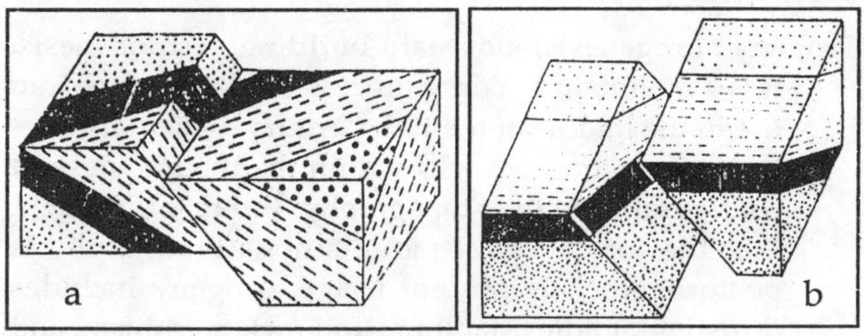

Fig. 103 a, b.

Oblique joint *(diagonal joint).*

Oblique-slip fault A fault on which the slip is oblique to the dip of the rocks *(diagonal-slip fault).*

Ocean-floor spreading *(sea-floor spreading).*

Oceanic ridge *(mid-oceanic ridge).*

Offset [fault] : normal horizontal separation In a fault, the horizontal component of displacement, measured perpendicular to the horizon.

Offset ridge A ridge that is discontinuous due to faulting.

Onlap It is *an overlap* characterized by the regular and progressive pinching out toward the margins or shores of a basin of the sedimentary sequences of rocks.

Open fold A fold in which the inter-limb angle varies from70° to 120° (Fleuty, 1964).

Orientation diagram It is a general term for a fabric diagram.

Orocline geoflex An orogenic belt with an imposed curvature or sharp bend. According to *Carey (1958)* due to horizontal bending the curvature is formed in the crust.

Orogen *(orogenic belt).*

Orogenesis *(orogeny).*

Orogeny : orogenesis; mountain building; tectnogenesis. It is a mountain building process. The term came into use in the middle of the 19th Century, which includes the deformation of rocks within the mountains, and configuration of their topography. Today, it is thought that the formation of mountainous topography is postorogenic. By present usage, orogeny includes thrusting , folding, faulting and metamorphism, and plutonism. Only in the late Cenozoic mountains there is an evident relation between rock structure and surfaces topography. All the deformational structures are products of orogreny.

Orogeosyncline Kober's term for a geosyncline, it is now regarded an area of orogeny (*Glassner & and Teichert, 1947*).

Orography orogeny; oreography. The physical geography dealing with the disposition, character, formation, and structure of mountains. It is the description of the relief of the Earth's surface or of a part of it, or is the representation of such relief on a map or model; the land features of a specified region. One of the orographic bends of the Himalayan mountain is the Western syntaxis near the Indus River. This syntaxial bend is thrust bound (Fig. 104, after Gansser, 1993).

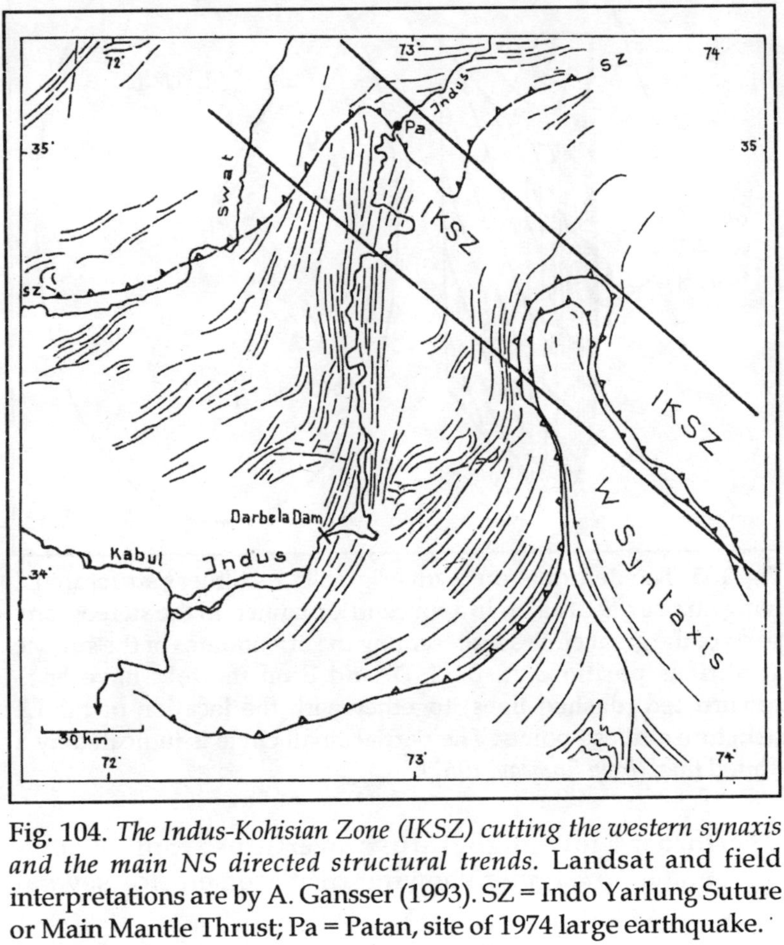

Fig. 104. *The Indus-Kohisian Zone (IKSZ) cutting the western synaxis and the main NS directed structural trends.* Landsat and field interpretations are by A. Gansser (1993). SZ = Indo Yarlung Suture or Main Mantle Thrust; Pa = Patan, site of 1974 large earthquake.

Orthotectonics (*Alpinotype tectonics*). *Dewey (1969)* used the trem for orogenic belts of the Andean type.

Overprint : superprint [struc geol] The superposition of a new set of structures on older ones like a *metamorphic overprint.*(Fig. 105).

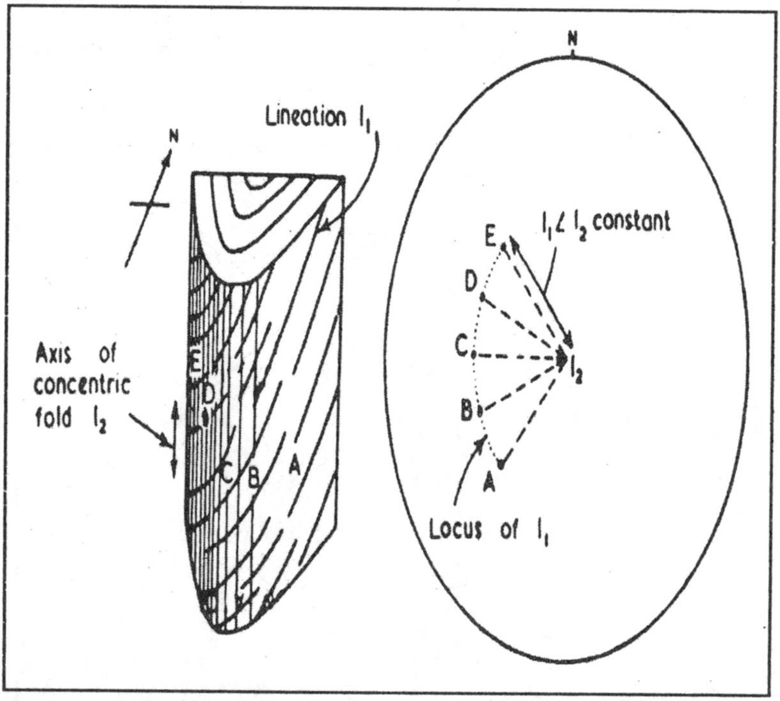

Fig. 105. The deformation of an originally rectilinear structure L1 lying on a surface folded in a concentric manner. In the stereogram, parts of the great circles representing the orientations of the surface ar various positions A, B, C, D, and E on the fold have been constructed (dashed lines) together with the location of the L1 structure at these points. The partial small circle is indicated by a dotted line. *(from Ramsay, 1967).*

Overthrust : low-angle thrust; overthrust fault A low-angle *thrust fault* showing displacement for several kilometers.

Overthrust nappe : overthrust block; overthrust sheet; overthrust slice. The hangingwall block of a large-scale overthrust (Fig. 106)

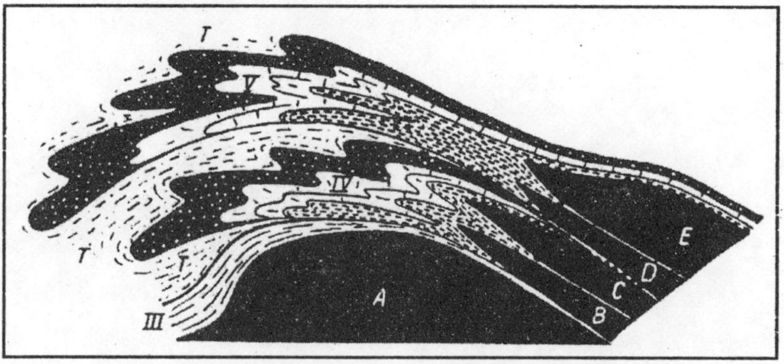

Fig. 106. Over thrust nappes and crystalline wedges A, Basement with sedimentary cover III; IV, V nappes with crystalline cores B, C, D, E; T, Teritiary between the nappes. Neritic calcareous facies, are solid black; Marl; shale and mudstone are broken lines. *(After Collet, in Hills, 1973).*

□□□

P

Palinspastic map A name proposed by Kay (1937) for a palaeogeographic map in which restored features are represented to their original geographic positions, prior to their shortening by folding, or thrusting. Saklani and their associates (1993) investigated the structure of the Garhwal Himalaya and revealed the geology along geological and restored/palinspastic sections (Fig. 107)

Parageosyncline : intrageosyncline A geosyncline within a craton or stable area of an epeirogenic origin (Stille, 1936). It is an oceanic depression marginal to the craton (Schuchert, 1923).

Paraliageosyncline A geosyncline of the present-day continental margin, (the Gulf Coast geosyncline).

Parallel fold: concentric fold A fold sharing constant orthogonal thickness of rocks (Fig. 108, 109).

Paratectonic An orogenic belt sharing steep cleavage in low-grade metamorphic rocks (Dewey, 1969).

Parautochthon Partially displaced rocks lying between *autochthons* and *allochthons.*

Parting : splitting A joint or fissure along which a rock is readily separated and divided into layers.

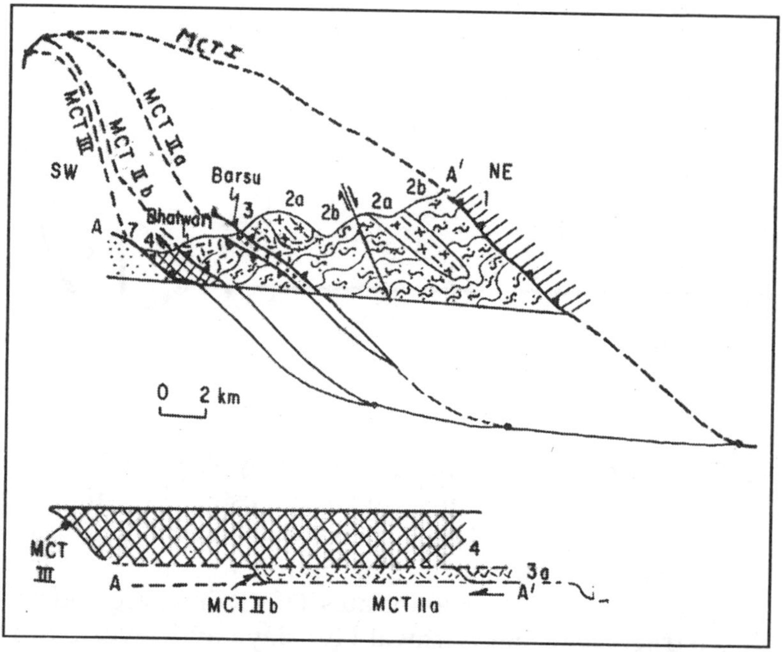

Fig. 107. Geological Cross-section and palinspastic (restored) section along line A-A. Kyanite-sillimanite gneisses and schist. 2. Biotite schist, 3. Porphyroblastic gneiss. 4. Amphibolitic migmatites, 5. Calcic migmatite, 6. Granite gneiss. 7. Quartzite (After Saklani, 1993).

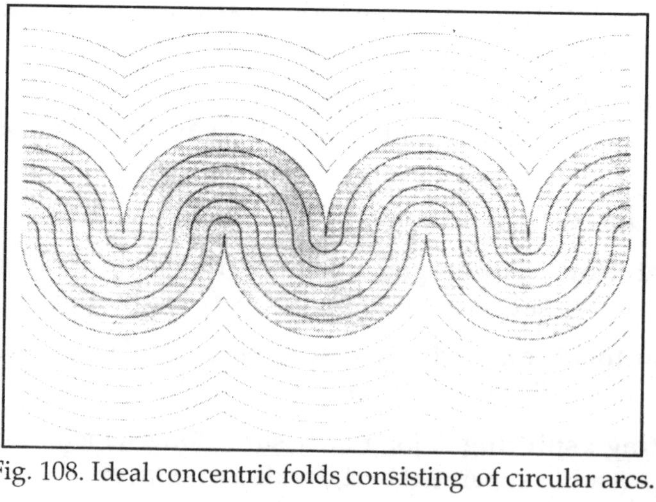

Fig. 108. Ideal concentric folds consisting of circular arcs.

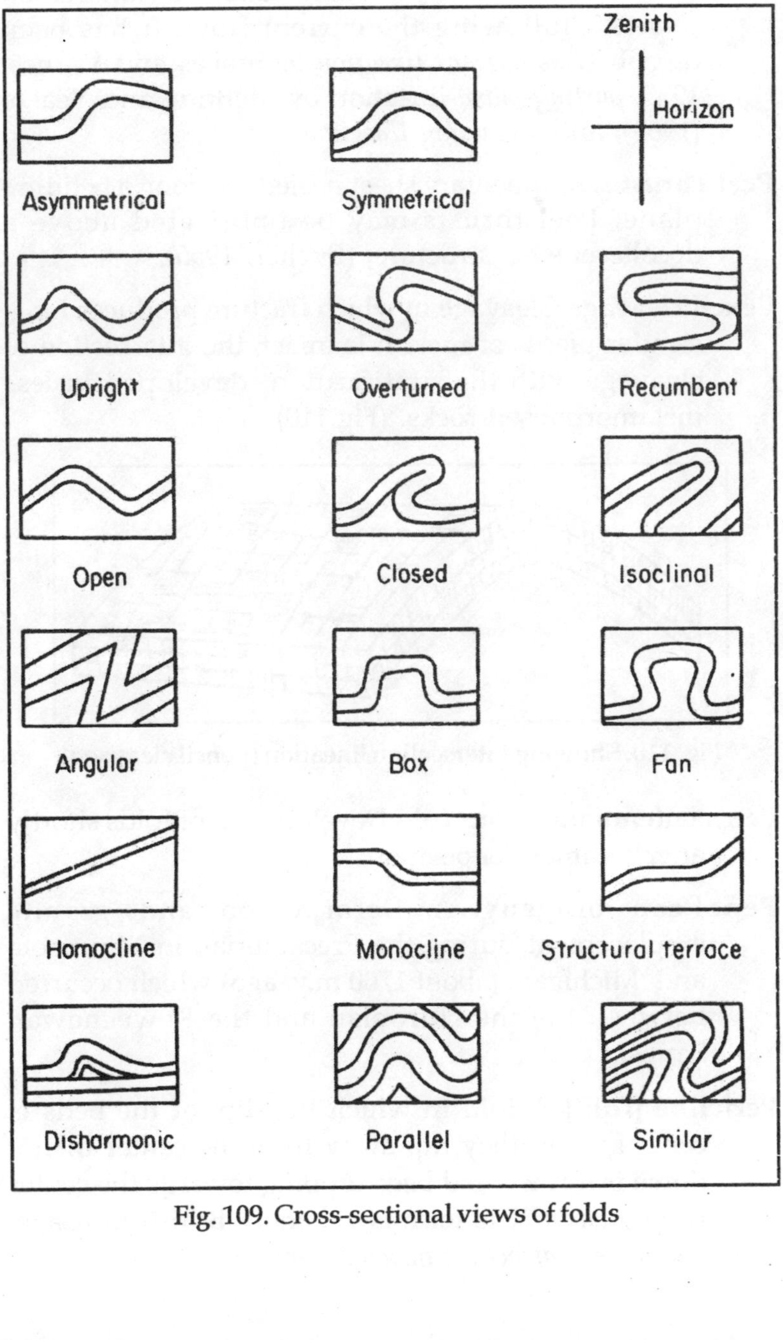

Fig. 109. Cross-sectional views of folds

Parting lineation It occurs as "faint streaks or lamination planes", following the current flow. It has been discribed as *current lineation by* Stokes and Varnes, 1955, *parting-plane* lineation by McBride and Yeakel (1963), and *streaming lineation.*

Peel Thrust A sedimentary sheet peeled off along a bedding plane. Peel thrusts may be imbricated above a décollecement structure. (Bucher, 1955).

Pencil cleavage Cleavage in which fracture produces long, slender pieces of rock. It is infact, the intersection of cleavage with the stratification developed in less metamorphosed rocks. (Fig.110)

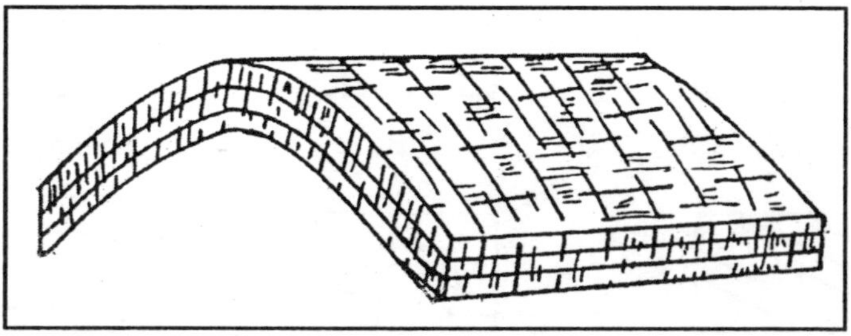

Fig. 110. Showing intersection lineation (pencil cleavage)

Penecontemporanceous fold Development of folds shortly after sediment deposition.

Penokean orogeny A Deformation and granite emplacement during the Precambrian in Minnrsota and Michigan, (about 1700 m.y. ago) which occurred between the the Huronian and the Keweenawan Series.

Pericline [fold] A fold in which the dip. of the beds is central when they dip away from the center then a dome is formed and beds dipping towards the center form a *basin.* The term is generally British in usage. See also: *centrocline; quaquaversal.*

Perpendicular throw The distance between two formerly adjacent points, measured perpendicular to the surface of the bed.

Petrofabric analysis It is used in *sturctural petrology.*

Petrofabric diagram *(fabric diagram).*

Pi axis *pi pole.*

Pillow breccia Anguler fragments of lava in a tuffaceous matrix .

Pinch-and-swell structure Quartz veins in metamorphosed rocks, show pinched and thinned structures due to deformation. (Ramberg, 1955) (Fig. 111).

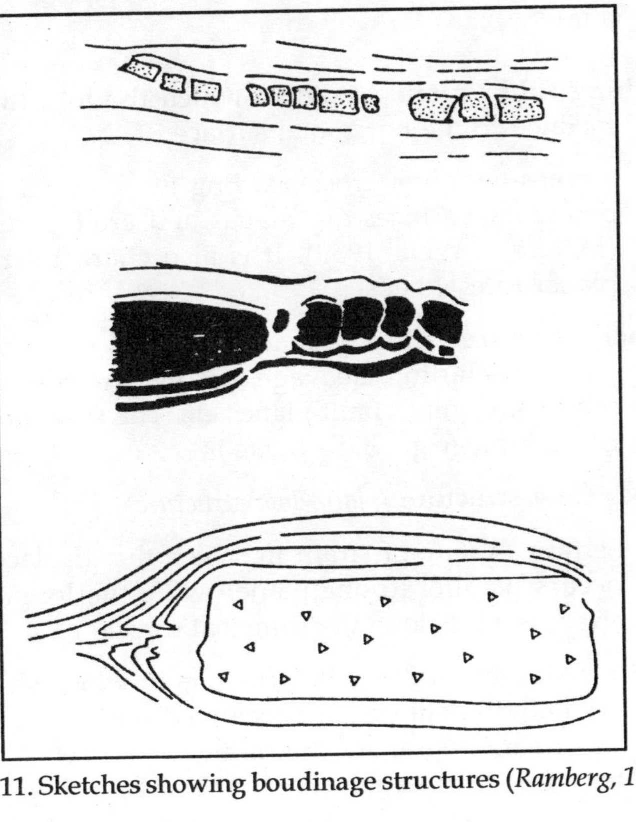

Fig. 111. Sketches showing boudinage structures (*Ramberg, 1955*).

Pi pole The pole, the normal to a fabric plane, is written as π pole. The β axis defined by the intersection of planes.

Pitch : rake The angle between the horizontal and the lineation, measured in the plane containing the lineation. (Fig. 112).

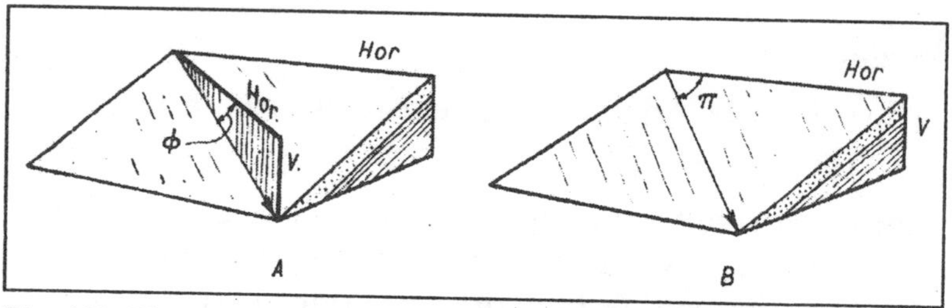

Fig. 112. Plunge and pitch of lineations A. ϕ is the plunge. B. π is the pitch or rake.

Pitching fold : pitch A fold in which the fold hinge is inclined to the horizontal surface.

Planar cross-bedding Cross-bedding in which the lower boundary surfaces are planar and eroded surfaces (McKee & Weir, 1953). It is also characterized by planar foreset beds.

Planar strucutres Closely spaced parallel mesoscopic/ planes including cleavage, distinct from foliation, schistosity, joint , fault planes etc. The structures can be multiple (e.g., S_1, S_2, S_3 etc.)

Planar flow structure *(platy flow structure).*

Plane strain A state of strain in which the displacement occurs parallel to one plane, while the longitudial strain is zero along the principal direction.

Plastic deformation It is a deformation in which shape or volume of a substance is changed without rupture. deformation slip, and twinning. It is a reheological term for deformation characterized by a *yield stress.*

Plate boundary : plate juncture; plate margin Seismic and tectonic zone along the edge of lithospheric plates, characterized by relative motion between plates (Dennis & Atwater, 1974)

Plate margin *(Plate boundary).*

Plate tectonics According to it the lithosphere is divided into a number of *plates* whose pattern of horizontal movement is that of torsionally rigid bodies intersecting one another at their boundaries, causing seismic and tectonic activity. (Fig.113)

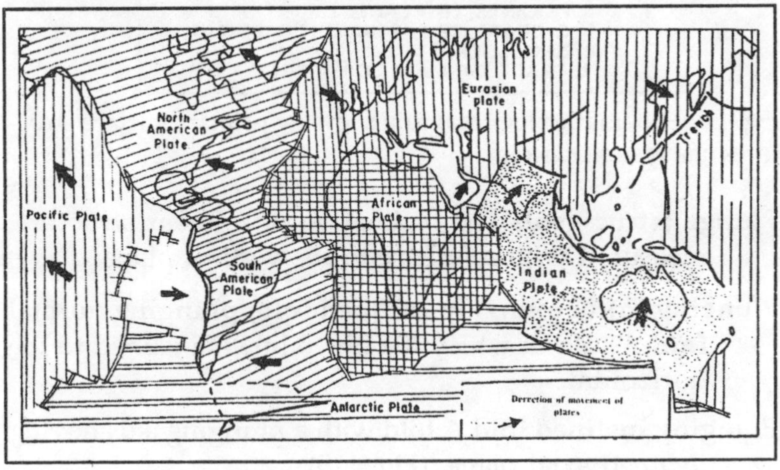

Fig. 113. Major plates of the world (after Valdiya, 1984).

Platform [tect] A platform is a part of the craton and consists of flat-lying or gently tilted strata, mainly sedimentary, underlain by a *basement* of rocks that were consolidated during earlier deformations.

Pilcation crinkled (small-scale folding).

Plume structure : plumose structure, perferred feather fracture A plumelike pattern, usually oriented parallel to the upper and lower surfaces of the containing rock (Fig. 114).

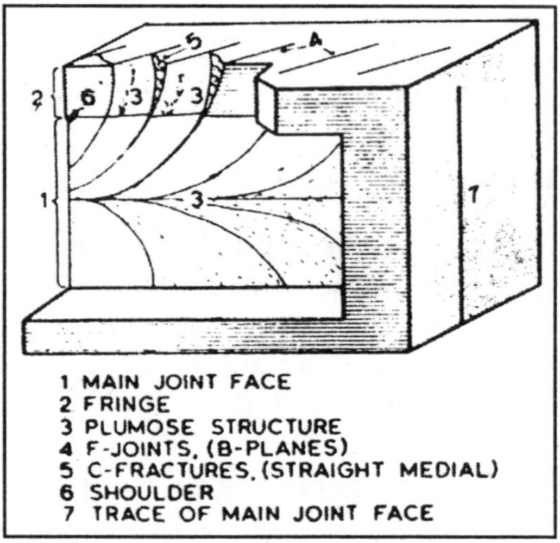

Fig. 114. Schematic block diagram showing primary surface structures of a systematic joint.

Plunge [struc geol] The inclination of a linear structure, measured with reference to the vertical plane.

Plunging fold : plunging inclined fold; plunging normal fold A fold of which the hinge line is inclined to the horizontal.

Plunging inclined fold A fold with a plunging axis having inclined axial plane. (Fig. 115)

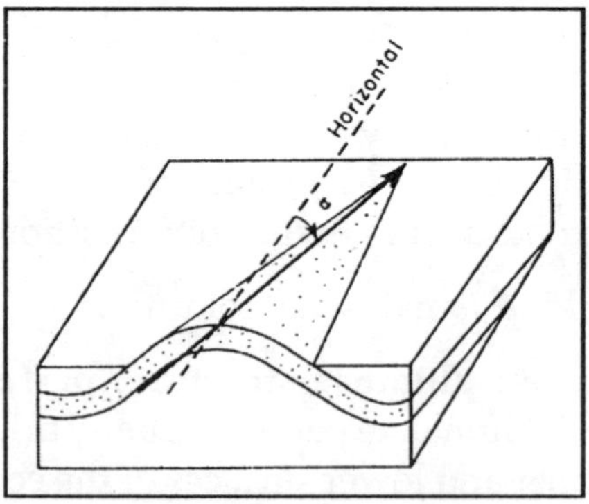

Fig. 115. Open, symmetrical, conical anticline (α = plunge)

Poisson's ratio It is a ratio of the lateral unit of strain with reference to longitudinal strain in a body stressed longitudinally within its elastic limit.

Polar wandering More or less systematic displacement of the Earth's poles, which possibly occurred during the geological time. (*polar migration; Chandler motion*).

Porphyroclastic structure (*mortar structure*).

Preferred orientation Orientation of planar or linear fabric elements, including crystallographic direction.

Primary fabric (*apposition fabric*).

Primary orogeny Orgeny that is characteristic of the *internides* and involves deformation, regional metamorphism, and granitization.

Primary structure [geol] (a) A structure in an igneous rock that originated contemporaneously with the formation or emplacement of the rock, but before its final consolidation; e.g. pillow structure developed during the eruption of a lava, or layering developed during solidification of a magma. (b) *primary sedimentary structure*, e.g. bedding or ripple marks. (c) The structure pre-existing the deformation and re-equilibration associated with the emplacement at shallow depth of a metamorphic massifs). (Fig. 116)

Principal axis of strain axis; principal axis. One of the three mutually perpendicular axes which were also mutually perpendicular before the deformation. There axes also represent the axes of the strain ellipsoid. The longest or greatest is the axis of elongation, the shortest or least is the axis of shortening.

Principal axis of stress axis; principal axis. One of the three mutually perpendiclar axes that are perpendicular to the *principal stress* (Fig. 117).

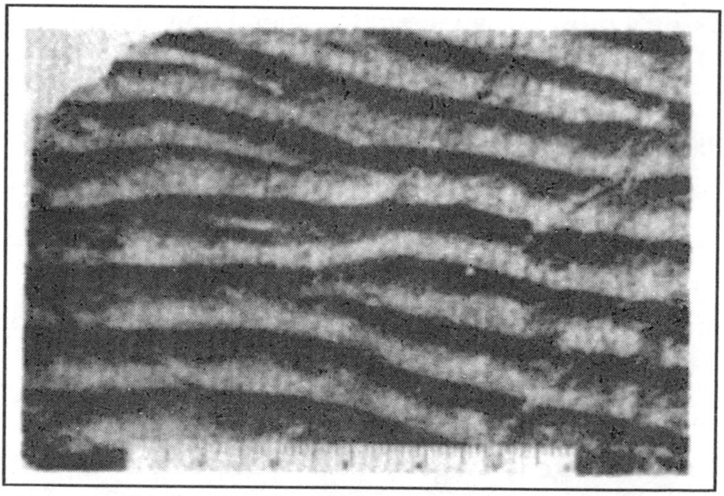

Fig. 116. Wave (or oscillation) Ripple mark. Cathedra Sandstone, Chapel Hill, Taggerty, Victoria. This photograph is of the upper surface of a bed; but owing to the common optical illusion or inversion of relief, it may appear as a cast. (In Hills, 1973).

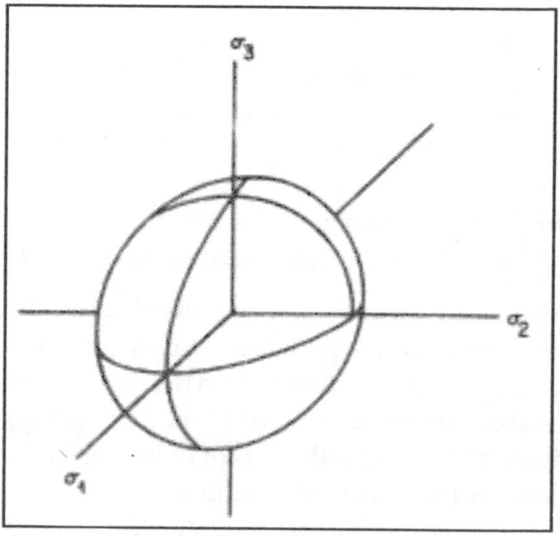

Fig. 117. Stress ellipsoid showing principal stress axes.

Prismatic structure *(columnar jointing).*

Prolate Elongated in the direction of a line of the poles .The equatorial diameters are much shorter than the dimensions from pole to pole (Fig. 118).

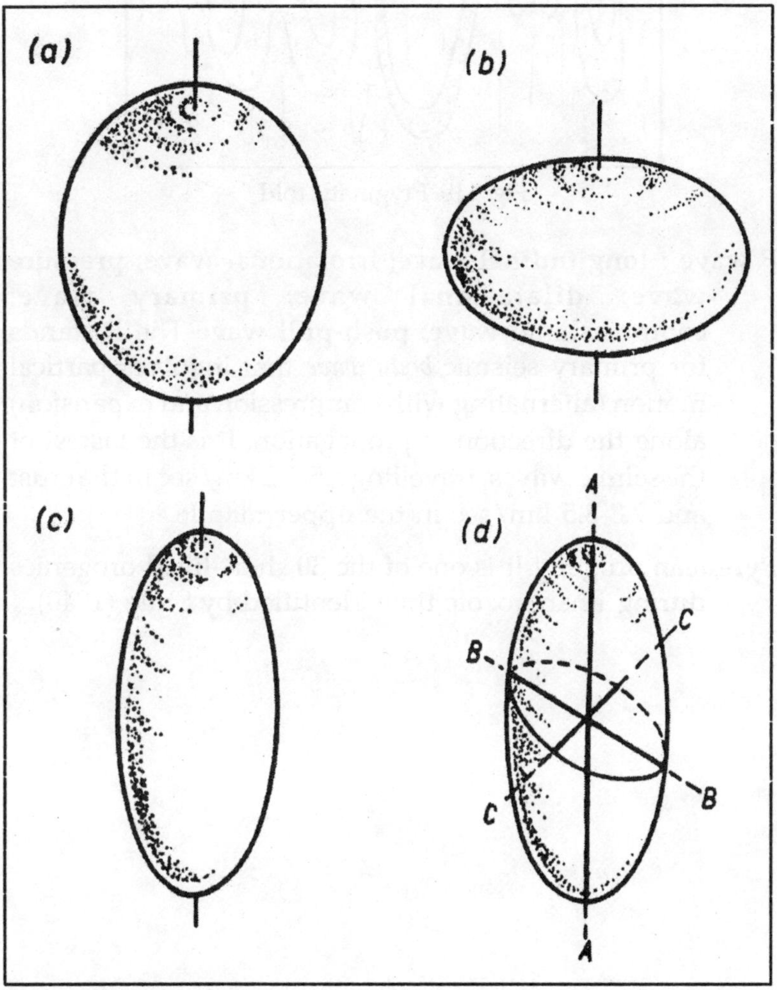

Fig. 118. The Strain Ellipsoid (a) Original shpere; (b) Biaxial ellipsoid (oblate); (c) Triaxial ellipsoid (prolate); (d) Triaxial ellipsoid.

Ptygma : ptygmatic fold Granitic material appering as disharmonic folds. (Fig. 119).

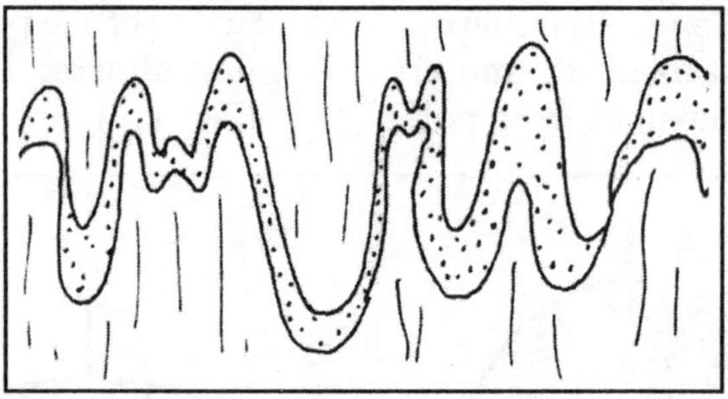

Fig. 119. Ptygmatic fold

P wave : longitudinal wave; irrotational wave; pressure wave; dilatational wave; primary wave; compressional wave; push-pull wave The P stands for primary seismic *body wave* that involves partical motion (alternating with compression and expansion) along the direction of propagation. It is the fastest of the seimic waves, travelling 5.5-7.2 km/sec in the crust and 7.8-8.5 km/sec in the upper mantle.

Pyrenean orogeny It is one of the 30 short-lived orogenies during Phanerozoic time identified by Stille (1940).

❑❑❑

Q

Quaquaversal : periclinal The term is used for structures that dip outward in all directions away from a central point. It is also called as *a* dome or ridge showing quaquaversal dips.

Quaquaversal fold *(dome).*

R

Radial fault The faults which radiate from a central point. (Fig. 120).

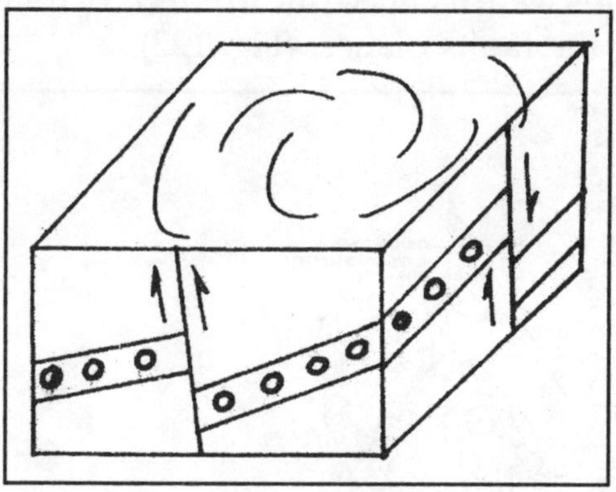

Fig. 120. Radial fault.

Ramp The steep and inclined segment of a thrust fault, formed due to imbrication/décollement changes occurring from lower to higher *stratigraphic levels.*

Rayleigh wave : R wave A *surface wave* with retrograde, and elliptical motion. It is named after Lord Rayleigh (Fig. 121).

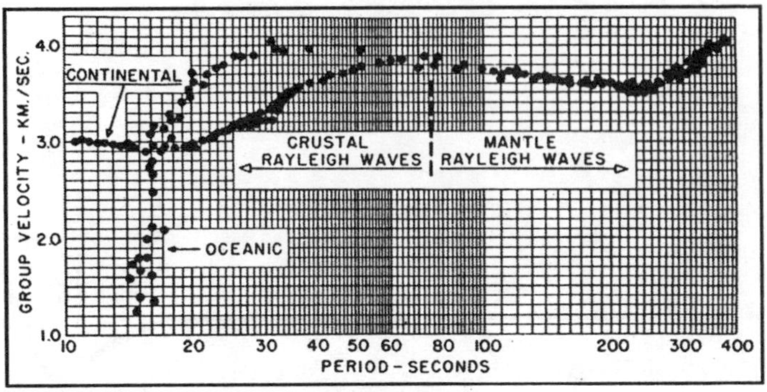

Fig. 121. Observed dispersion of Rayleigh waves in the period range 10 to 400 sec.

Reclined fold A fold in which hingeline plunges parallel to the dip direction of the axial surface (Turner & Weiss, 1963). Shekhar et al., (2006) described the vertical folds near the North Almora Thrust which are of reclined origin. During the Himalayan orogeny the folds were reclined due to effects of rotational stress along the faults planes. (Fig. 122).

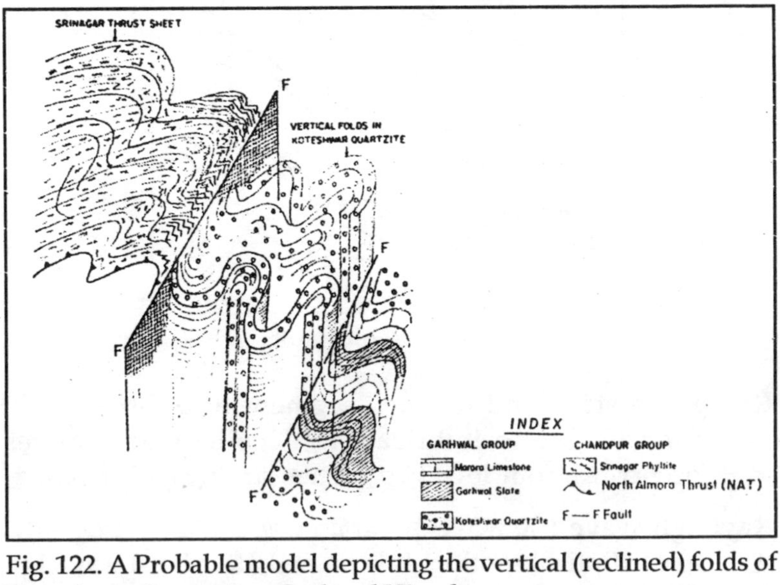

Fig. 122. A Probable model depicting the vertical (reclined) folds of Koteshwar Quartzites, Garhwal Himalaya.

Recumbent fold An overturned fold, having nearly or horizontal axial plane (Turner & Weiss, 1963). Schwan and Saklani, (1991) described their occurrence in rocks of Dunda area, Garhwal Himalaya (Fig. 123).

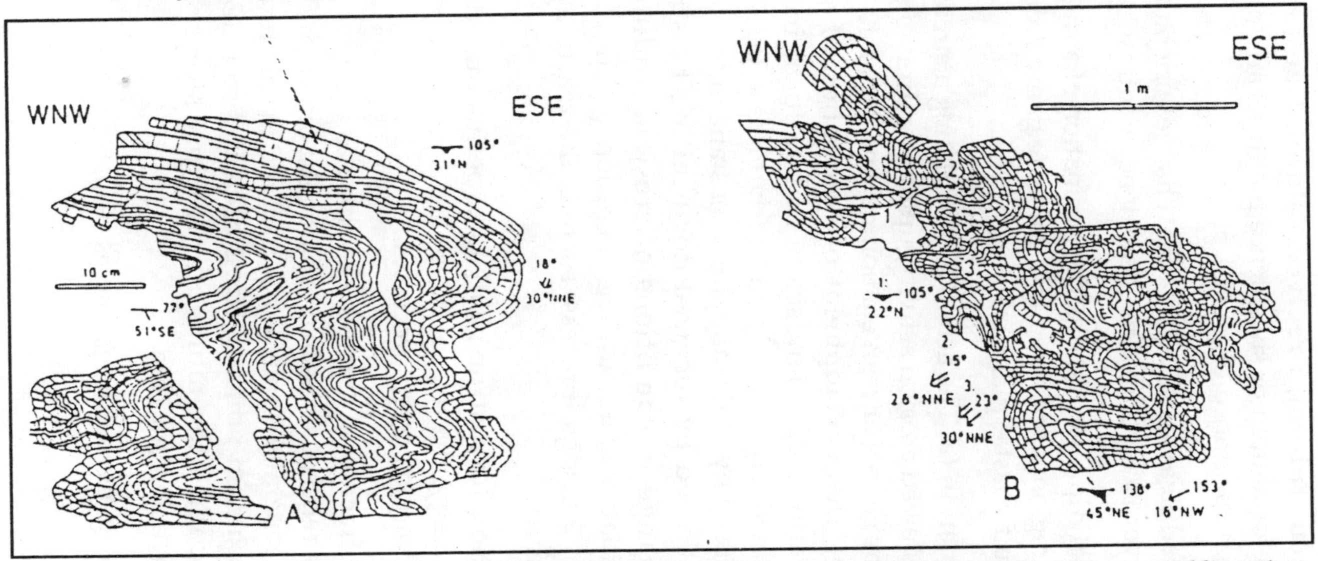

Fig. 123. (A+B) Recumbent folds with dominantly Himalayan strike. The NE to N scattering of axes and lineations are also common in the ruby-coloured Bhainga slates intercalated with quartzitic beds in the Dunda Window : 4.5km south of Dunda. (After Schwan and Saklani, 1991).

Recurrent folding : revived folding Due to periodic deformatiom or subsidence the rocks are folded with disappearance of crestal parts.

Refracted cleavage Cleavage that changes orientation from layer to layer due to competency contrast.

Regional dip : normal dip The inclination of strata over on a regional scale at a low angle, (e.g. the Atlantic and Gulf coastal plains).

Rejuvenation (a) The action of stimulating a stream due to renewed erosion activity, uplift or a drop of sea level. The causes of rejuvenation may be dynamic, or static. (b) The development or restoration of youthful features of a landscape or landform in an area previously worn down nearly to base level, usually caused by regional uplift or eustatic movements. It is followed by renewed downcutting by streams; a change in conditions of erosion, leading to the initiation of a new cycle of erosion. (c) The renewal of any geologic process, such as the reactivation of a fissure.

Relaxation [exp struc geol] In experimental structural geology, the release of applied stress with time, due to various creep processes.

Release fracture A fracture develped as a consequence of the relief of stress in one particular direction. The term is generally applied to a fracture. In this process the maximum principal stress decreases sufficiently and becomes the minimun principal stress. It is an extension fracture.

Residual geosyncline *(autogeosyncline).*

Reversed *(Overturned).*

Reversed fault *(reverse fault thrust fault reversed fault; reverse silp fault.)*

Reverse fault A fault in which hangingwall block goes up relative to the footwall block. (Fig. 124).

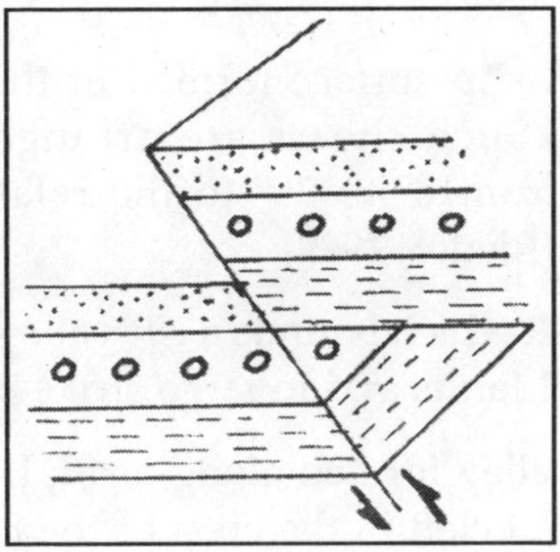

Fig. 124. Reverse fault

Reverse similar fold A fold showing axial region with thickened limbs.

Rheid : fold rheomorphic fold A folded strata deformed by flow (Fig. 125).

Fig. 125. Rheid fold.

Richter scale A term proposed by seismologist C.F. Richter. Infact there is no upper limit to the magnitude of an earthquake, but the strength of Earth materials produces an actual upper limit which is slightly less than 9.

Riedel shear A slip surface formed at the early stage of shearing. Such shears are arranged en échelon, inclined from10° to 30° to the relative movement direction (Riedel, 1929).

Rift fault A fault which bounds a rift valley. The term used for normal faults and to large strike-slip faults.

Rift valley A valley formed along a *rift.* It also represents deep central cleft in the crest of the *mid-oceanic ridge.*

Right-lateral slip fault *(right-lateral fault).*

Rigidity The property which resists the applied stress which would distort it. A fluid is devoid of rigidity.

Rim syncline : peripheral sink In salt tectonics. a local depression is developed at the borders around a salt dome.

Ring fault : ring fracture A steep-sided cylindrical fault pattern associated with *cauldron subsidence.*

Ring fracture *(ring fault).*

Ripple bedding (a) A term used by Hills (1973) to *current bedding* for "the small-scale ripple-like bedding in sand".

Rock flowage *(flow).*

Rodding In metamorphic rocks, a linear structure in which the stronger parts, such as veined quartz or quartz pebbles, are oriented parallel to rods. Its origin is debatable. (Fig.126).

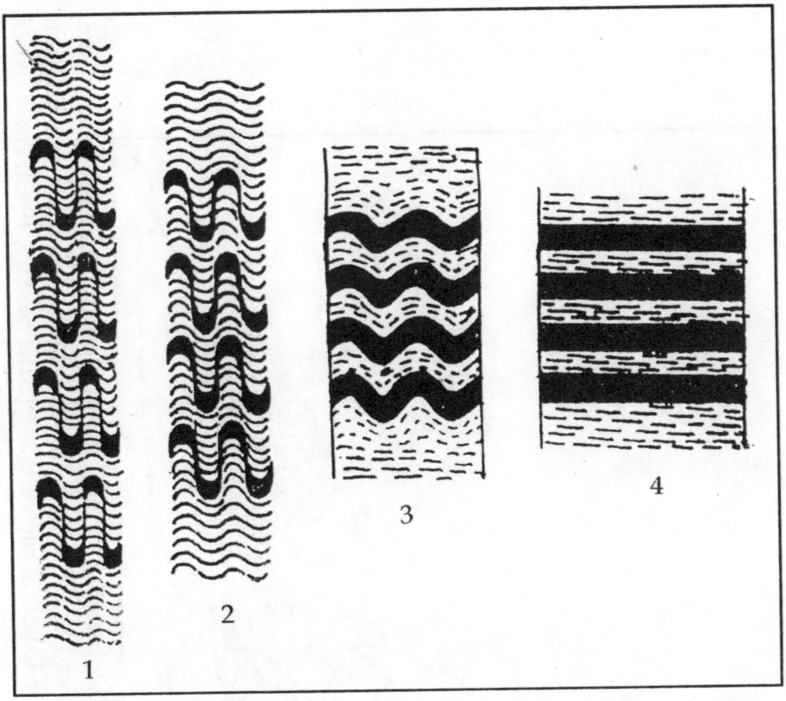

Fig. 126. Various stages for the development of Rodding (1 to 4).

Roof thrust The upper boundary of a *duplex structure* (Dahlstrom, 1970) (Fig. 127).

Root [fold] The basal part of a nappe that was originally linked to its source, or *root zone*.

Root zone [fold] root scar The source or original place of the *root* of a nappe.

Rotation : cylindroidal fold A cylindrical fold, whose axial surface is rotated or destroyed by subsequent cross - folding (Whitten, 1972).

Rotational fault A fault which shows rotation of displaced rocks.

Rotational fault A fault on which the rocks are *rotated*.

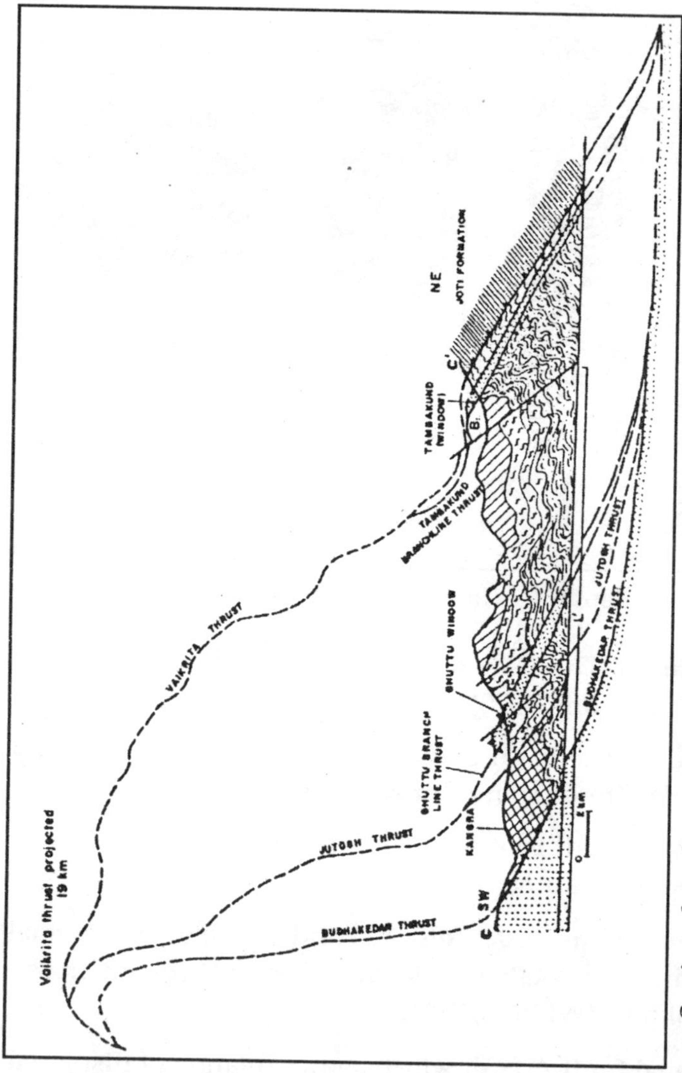

Fig.127. Cross Section along line C-C″ Current Duplex Length (L′) 26.6km. Structural Thickness (H′) 19.4 km. Stratigraphic Thickness (t′) 8.5 km. Cross Section Area (A) 516 sq km. Initial Length of Duplex (Lo) 60.7 km. Current angle between Floor Thrust and central portion of subsiding faults 37°. (After Saklani, 1993). The Vaikrita Thrust is the roof thrust.

Rule of V's The outcrop or a rock-bed when crosses a valley then V is formed in the direction in which the outcrop lies underneath the stream. The V points upstream where the ourcrops of horizontal beds parallel the topographic contours; where the beds dip upstream, or where the beds dip downstream at a smaller angle than the gradient of the stream; the V points downstream where the rock-beds dip downstream at a larger angle than the stream gradient. (Figures 128, 129, 130, 131 and 132; After Lahee, 1971).

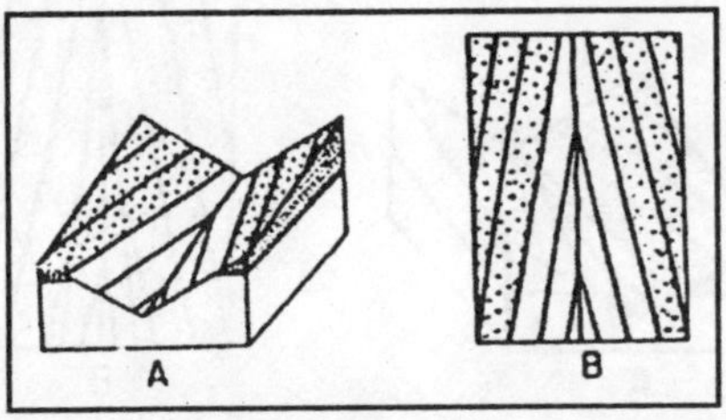

Fig. 128. Relations of horizontal bedding to contours shown in a block diagram (A) and in a map of the surface of the block (B) Stippled and blank portions represent two rock strata.

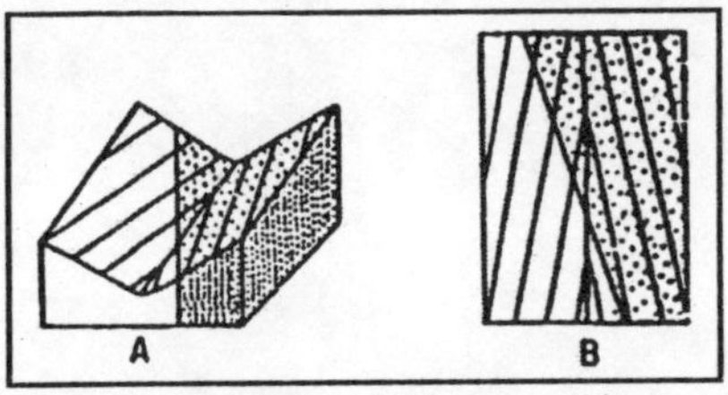

Fig. 129. Relations of vertical strata to contours

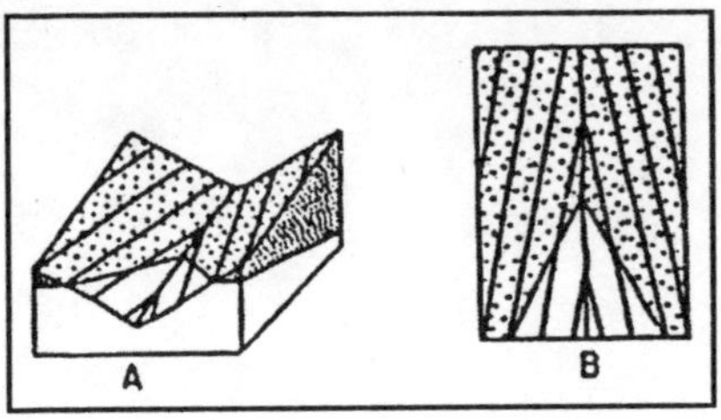

Fig. 130. Relations of inclined strata to contours.

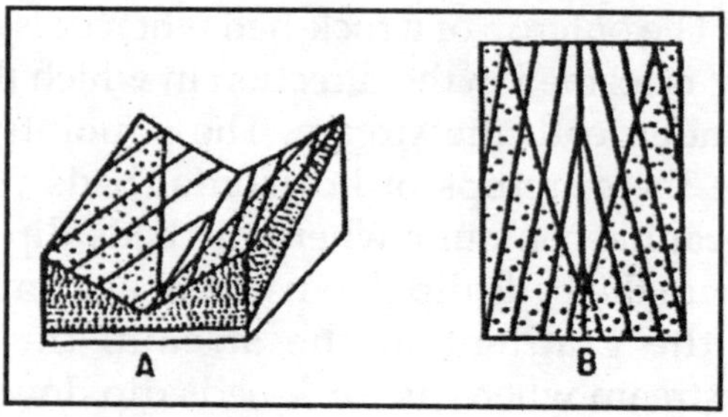

Fig. 131. Relations of inclined strata to contours.

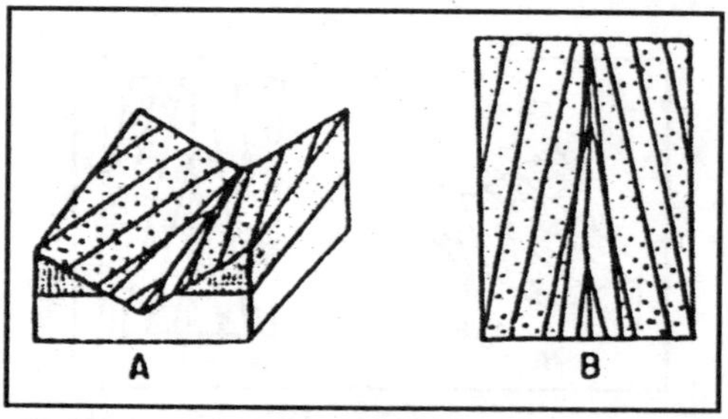

Fig. 132. Relations of inclined strata to contours.

S

Saddle A depression located along the axial trend of an anticline.

Sag structure (sedimentary) A term used for load casts and related structures.

Salt anticline : salt wall A diapiric structure, like a *salt dome,* the core of which is linear rather than equidimensional, e.g., the salt anticlines in the Colorado Plateau.

Salt dome A *diapiric* structure having an equdimensional *salt plug,* measuring from one to two kilometers in diameter, in the enclosing sediments from a mother salt bed 5 km to more than 10 km beneath the top of the plug. The salt plugs have a *cap rock* consisting of anhydrite and have nearly vertical walls, Commonly the enclosing sediments are complexly faulted. The permeable beds bear oil and gas. Salt domes are characteristic features of the Gulf Coast America and Germany. (Nettleton 1934). (Fig. 133).

Salt-dome breccia This type of breccia occurs in a shale sequence of a dome-shaped mass surrounding a salt plug and is formed by differential pressure of diapiric intrusions (Kerr & Kopp, 1958).

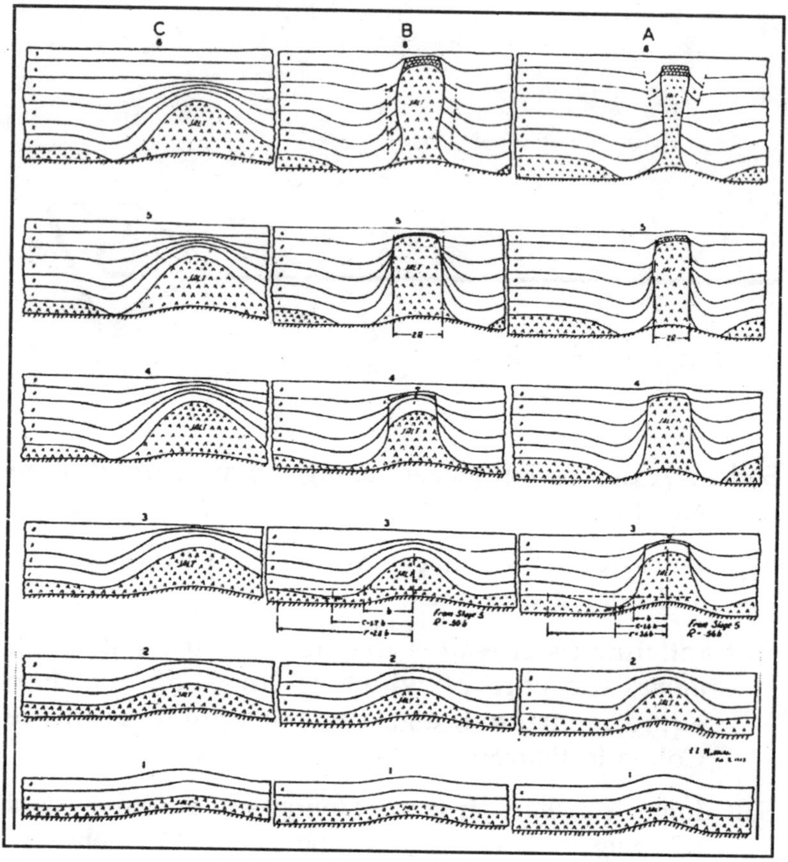

Fig. 133. Formation of salt dome structure in scale-model experiments. (*From Nettleton, 1934*).

Salt stock A diapiric salt body of different shape.

Sandy breccia A breccia containing 80% rubble, 10% sand, and 10% of other material.

Savic orogeny One of the 30 or more short-lived orogenies during Phanerozoic time identified by Stille. (1940)

Scarp fault : escarpment scarp; erosion scarp A cliff produced by faulting or by erosion and due to differential movements of the landslides.

Schistosity It is a foliation defined by parallel, platy, prismatic, or ellipsoidal minerals in a coarse grained rock. Commonly the foliation consists of micaceous minerals. (Fig. 134)

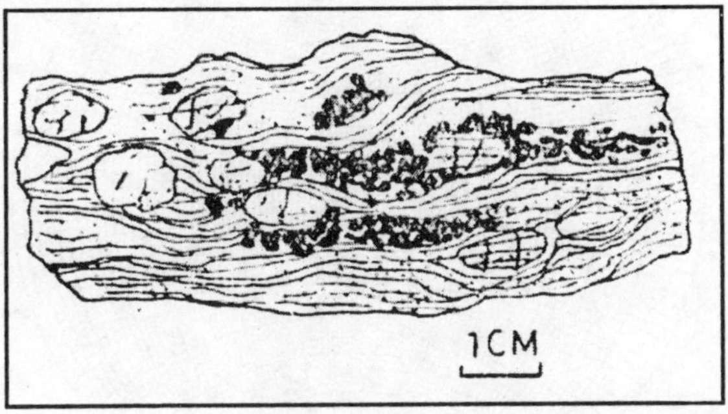

Fig. 134. A schist containing mica flakes, quartz and tourmaline.

Schmidt net It is a coordinate system to plot a *Schimdt projection,* used for statistical analysis of data obtained. from universal-stage measurments and from field for plotting the planar and linear structures (Fig. 135).

Schuppen structure (*imbricate structure) (*German *Schuppenstrucktur).* When rocks are folded and are faulted with same angles of dip along parallel faults.

Scissor fault It is a fault which shows increasing offset or separation along the strike. A scissorlike or pivotal movement on the fault, is characterized by uniform strike-slip movement across a synclinal or anticlinal fold. (*pivotal fault, hinge fault, rotational fault, differential fault).*

Scour cast A sole mark consisting of *flute cast of sadimentary origin.*

Sea-floor spreading : ocean-floor spreading ; spreading concept; spreading-floor hypothesis The rise of the oceanic crust is caused by convective upwelling of magma along the *mid-oceanic ridges.* The spreading

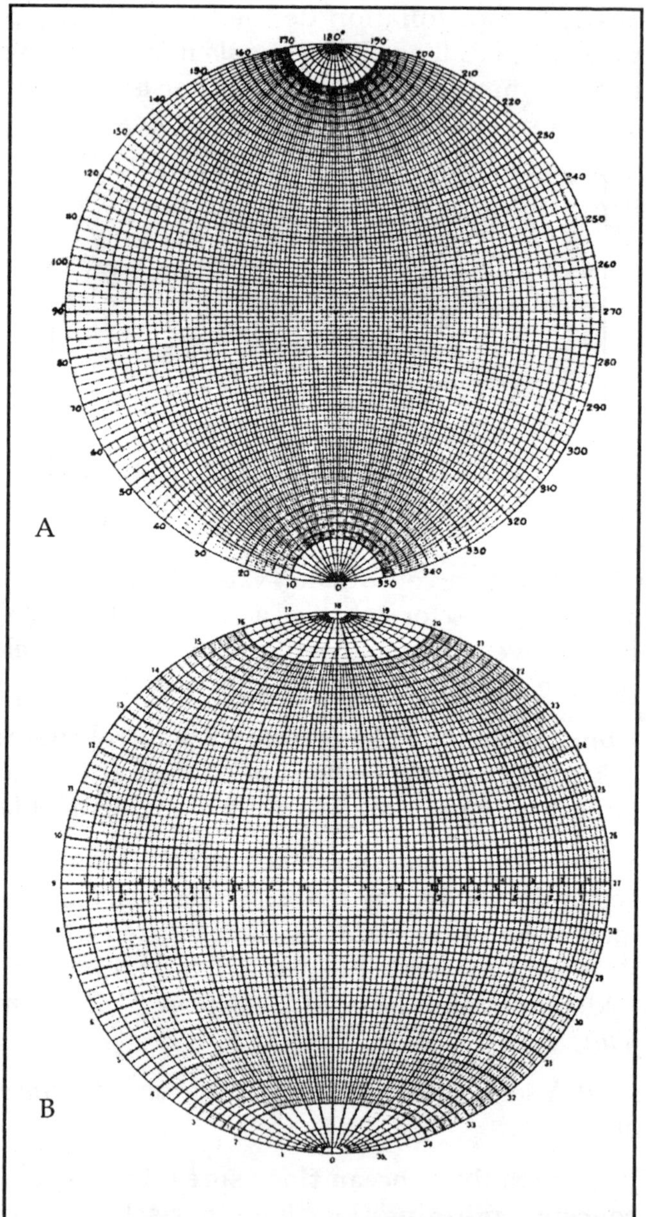

Fig. 135. A. Stereographic or Wulff net.
B. Lambert's equal area net, or Schmidt net.

away of the new material at a rate of one to ten centimeters per year.

Secondary cleavage It is a type of foliation plane after deformation and /or metamorphism.

Secondary creep : steady-state creep It is a slow deformation of material under a constant differential stress having strain as a constant.

Secondary structure A structure formed subsequent to the deposition of the rock in which it is found, such as a fault, (e.g. epigenetic *sedimentary structure,* such as a concretion or nodule), or a sedimentary dike.

Secondary tectogenesis Gravitational sliding of a tectogen and the resulting deformation (Haarmann, 1930). It is now an obsolete term.

Sedimentry breccia : sharpstone conglomerate This *breccia* is formed by a sedimentary process where the rock materil are jumbled together e.g., a talus breccia. It is characterized by imperfect mechanical sorting (Wentworth 1935).

Sedimentary fault *(growth fault).*

Sedimentary structure This *structure* is present in a layered rock which is syn-sedimentational or it is a product of sedimentary processes subsequent to deposition.

Sedimentary tectonics When rock-strata are subjected folding and deformation in basins then a large anticline is developed at depth in a geosyncline (Gussow, 1962).

Seismic creep A slow deformation along a fault.

Seismic gap A tectonically active zone which has not experienced *earthquake* during a time interval. Seismic gaps are highly prone to earthquakes.

Shank [fold] It is an obsolete term used for *a limb* of a fold.

Shatter breccia : A breccia of tectonic origin which is composed of angular fragments showing very little rotation.

Shear It can be of two types i.e., simple and pure shear. The term refers to strain rather than to stress. It is also used for the zones of shear, failure of and to surfaces characterized by differential movement (Fig. 136).

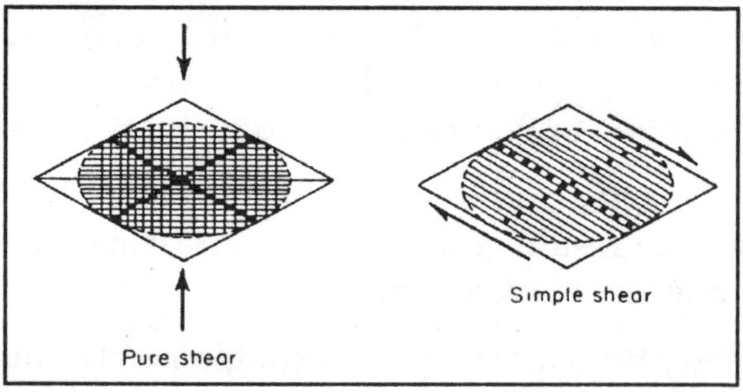

Fig. 136. Internal distinction between pure and simple shear.

Shear cleaveage : slip cleavage This term refers to cleavage characterized by displacement of the pre-existing surfaces across the cleavage plane (Fig. 137).

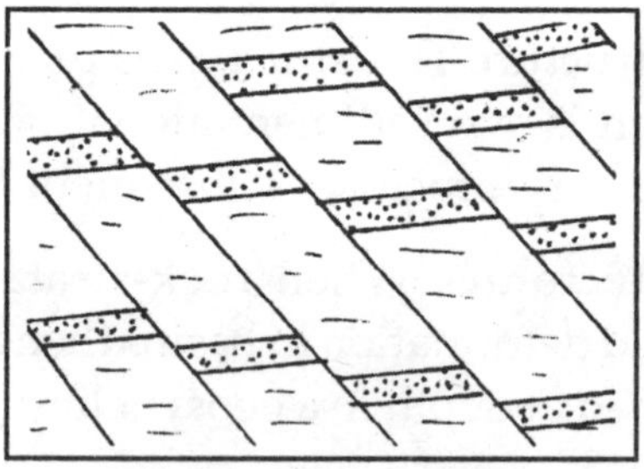

Fig. 137. Slip cleavage.

Shear fracture A fracture formed by stresses which tend to shear one part of a rock past the other.

Shear plane (*shear surface*).

Shear strain Due to this, the parallel lines are sheared past one another by deformation. It is the tensor measure that appears in *infinitesimal-strain theory.*

Shear stress The component of *stress* affecting tangentially the plane through any given point in a body.

Shear surface : shear plane A surface along which differential movement has occurred.

Sheet structure *(sheeting).*

Shield : continental shield ; cratogene; continental nucleus The area of *basement rocks* in a craton, commonly, surrounded by sediment-covered *platforms;* e.g., Canadian Shield, Baltic Shields etc. which are of Precambrian age.

Shift : slip The rock- displacement due to fault but outside the fault zone.

Shoulder A structure caused due to the intersection of *plume structure* with *fringe joints.*

Simple shear The strain characterzed by fixed orientation of one of the circular section of the *strain ellipsoid.* The process is like shearing a deck of cards in one direction.

Sinistral fault *(left-lateral fault).* (Fig. 138).

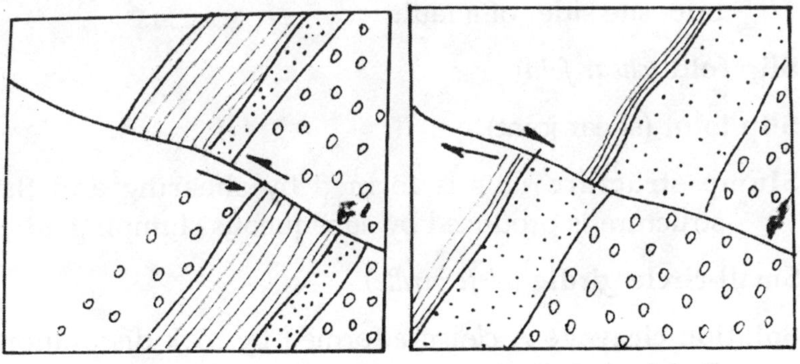

Left Lateral fault Right Lateral fault

Fig. 138.

Sinistral fold An asymmetric fold which is S-or Z-shaped (Fig. 139).

Z- shaped fold S- shaped fold

Fig. 139.

Slaty cleavage : flow cleavage A pervasive, parallel foliation of fine-grained, platy minerals (mainly chlorite and sericite) in a direction perpendicular to the direction of maximum finite shortening, developed in slate or other homogeneous sedimentary rock by deformation and low-grade of metamorphism. Mostly the slaty cleavage is also *axial-plane cleavage.*

Slice (*thrust slice).*

Slip : total displacement The relative displacement on opposite sides of a fault.

Slip fold *(shear fold).*

Slip joint (shear joint).

Slump structure This is formed by shearing and the structure is produced by subaqueous slumping.

Small-circle girdle *(cleft girdle).*

Solution cleavage A *cleavage* formed by rock dissolution, The stylolitic cleavage has about 10% shortening parallel to bedding". (Alvarez et al., 1978).

Spaced cleavage Cleavage charcterized by the spacing or separation of mesoscopic to microscopic planes (e.g. *slip cleavage).*

Sphenochasm It is a triangular gap of oceanic crust separated by two continential blocks converging to a point. According to *Carey (1958)* it is originated due to the rotation of one of the continental blocks with respect to the other.

Splaying out The breakup and splitting of a fault into many minor faults.

Spreading-floor hypothesis *(sea-floor spreading).*

S-tectonite A deformed rock whose fabric is dominated by planar structures (e.g., slate) (Fig. 140).

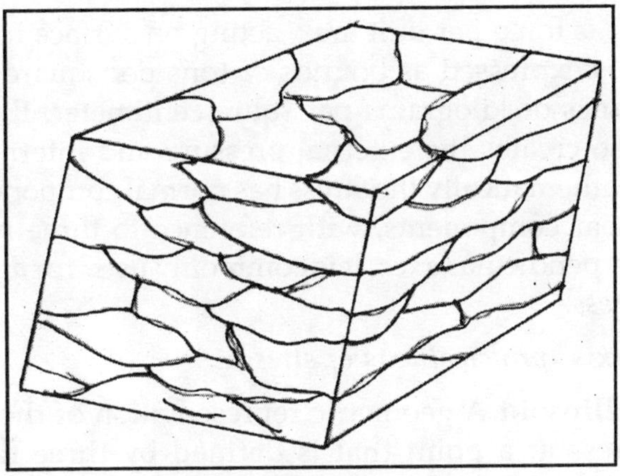

Fig.140. S-tectonite

Step fault (a) One of a set of parallel, closely spaced faults over which the total displacement is distributed. (b) One of a series of low-angle thrust faults in which the fault planes step down and laterally in the stratigraphic section to lower glide plane. Step faulting is due to variation in the competence of the beds in the stratigraphic section (Jones, 1971).

Step fold : monoclinal structure An abrupt downward flexure of horizontal strata.

Stereogram A diagram depicting sterographic projections of structural data on the net.

Stereonet *(Wulf net).*

Stick-slip A jerky, sliding motion associated with fault movement. It can be a mechanism of shallow-focus earthquakes.

Strain deformation A resultant of stress which brings a change in the shape or volume of a body .

Strain ellipse A geometric representation of strain defined by three mutually perpenlicalar principal axes.

Streaming lineation *(parting lineation).*

Stress It is force per unit area, acting on surface in a solid. It is expressed as pounds or tons per square inch, or dynes or kilograms per squre centimeter. Extension, also creates the external pressure and internal force. Mathematically the stress has normal components and shear components, with reference to three mutually perpendicular axes. It is commonly used for *differential stress.*

Stress axis *(principal axis of stress).*

Stress ellipsoid A geometric representation of the state of stress at a point that is defined by three mutually perpendicular principal stresses and their intersection.

Stretch According to Malvern (1969), it is a measure of change in length of a line, specifically the ratio of the final length to the intial length of the line.

Striation (a) One of multiple scratches or minute lines, generally parallel, inscribed on a rock surface by a geologic agent, i.e. glaciers or faulting; (b) The condition of being striated; the disposition of striations.

Strike The horizontal direction or trend of a structural surface, e.g. bedding or fault plane. It is at right angles to the dip.

Strike fault. A fault which is formed parallel to the strike of the strata.

Strike fold *(longitudinal fold).*

Strike joint A joint that strikes parallel to the strike or lineation in a rock.

Strike separation It is a *separation* of two formerly adjacent beds on either side of a fault surface and is measured parallel to the strike of a fault.

Strike-shift fault *(strike-slip fault).*

Strike-slip fault : strike-shift fault A fault showing the displacement parallel to the fault's strike.

Structural closure *(closure).*

Structural contour *(structure contour).*

Structural geology It deals with the form, internal structure of the rocks, and minerals and their analysis on mesoscopic scale. It is similar to *tectonics,* but the latter is used in a broader sense.

Structural petrology : fabric analysis; petrofabric analysis; petrofabrics; microtectonics Study of rock-fabric on micro scale. It includes relationships of (microstructure) and the preferred orientations of minerals and their deformation. The crystal structures are also studied under Electron microscope.

Structural trap A *trap* for oil or gas formed as a result of folding, faulting, or other deformation.

Structural [struc geol] (a) The general disposition, attitude, arrangement, or relative positions of the rock masses of a region or area; the sum total of the *structural features* of an area, consequent upon such deformational processes as faulting, folding, and igneous intrusion. (b) A term used in petroleum geology for any physical arrangement of rocks (such as an anticline or reef) that may hold an accumulation of oil or gas.

Structural contour : subsurface contour A *contour* that protrays a structural surface consisting of boundary or faults.

Structural contour map : structure map A map depicting subsurface configuration with the help of *structure contour* lines.

Subduction In this process, one lithospheric *plate* descends beneath the another plate. A related concept was originally used by Alpine geologists. See also: *subduction zone* (Fig. 141).

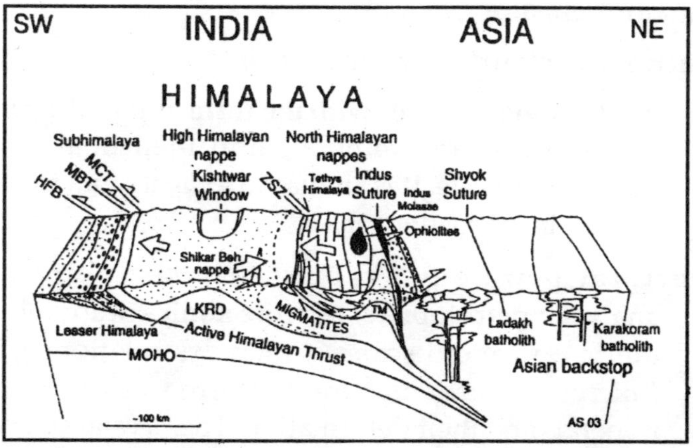

Fig. 141. The main tectonic units of the NW Indian Himalayan HFB = Himlayan Frontal Boundary, LKRD = Larji-Kallu-Rampur dome MBT = Main Boundary Thrust, MCT = Main Central Thrust, ZSZ = Zanskar Shear Zone. Indus suture is the subduction on zone (After Steck, 2003).

Subfabric Arrangement of one kind of structural element in a rock.

Subsidence The sinking or downward settling of the Earth's surface with little or no horizontal motion. It can be caused by geologic processes, such as solution, crustal warping, and as subsurface mining, *land subsidence; bottom subsidence.* Subsidence can occur due to the formtion of a rift valley or lowering of coastal margins due to tectonics.

Subsidiary fold *(parasitic fold).*

Superficial surficial It pertains to a surface or surface layer or a "superficial structure"formed by surface creep.

Superficial fold *(décollement fold).*

Superposed fold (*Superimposed fold).*

Superstructure : supra structure The upper structural layer in an orogenic belt, subjected to relatively shallow or near-surface deformation in contrast to an underlying and more complexly deformed metamorphosed *infrastructure.*

Symmetrical fold : normal fold A fold sharing same angle of dip of the limbs relative to the axial surface. (Fig. 142).

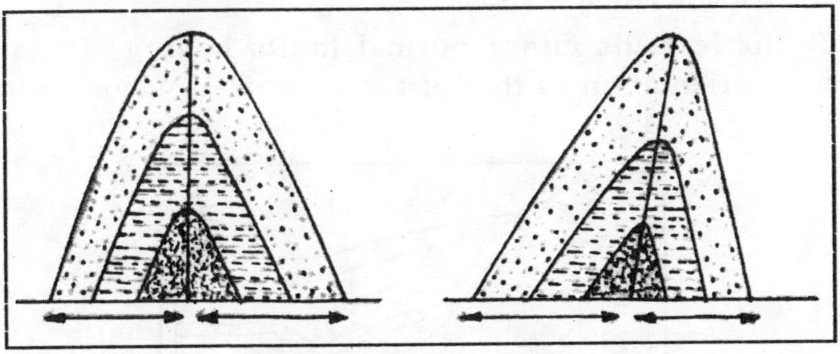

Fig. 142. Symmetrical fold

Symmetry All the elements making up the fabric is collectively knwon as symmctry It is of five types (1) spherical (2) axial, (3) orthorthombic (4) monoclinic and (5) triclinic, (Turner and Weiss, 1963).

Symmetry axis ; rotation axis An imaginary line about which the crystal may be rotated, with representions of lines, angles, or faces. It is one of the *symmetry elements.*

Symmetry elements Crystal symmetry can consist of axes, plane, and center of symmetry described. There are 32 possible classes of symmetry; each arrangement is a crystal *class.*

Synclinal A fold whose/core is constituted by stratigraphically younger rocks.

Synclinorium A major synclinal structure of regional scale having many folds.

Syntaxis Due to tectonic movements the rocks acquire a sharp bend in an orogenic belt.

Syntaxy Similar crystallographic orientation in a mineral grain and its overgrowth.

Syntectonic : synorogenic It is a geologic process or episods during a tectonic activity affecting the rocks.

Syntexis Magma formation due to rock melting and/or assimilation of rocks.

Sythetic fault minor normal faults having similar orientation as that of the associated major fault. (Fig. 143).

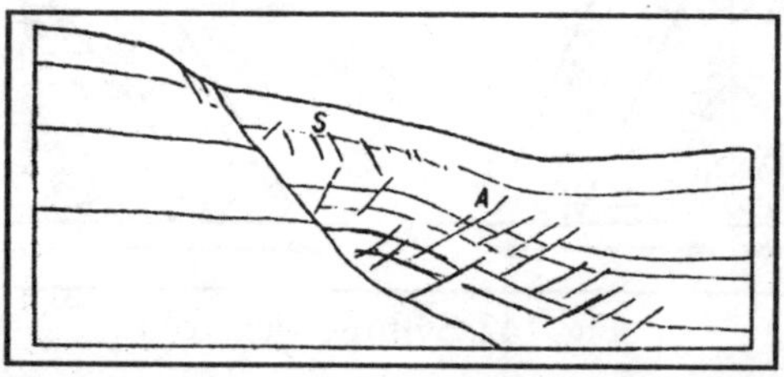

Fig. 143. Antithectic (A) and Synthetics faults along the edge of an artificial graben in clay.

❑❑❑

T

Tacnian orogeny : Taconian orogeny An orogeny of the Ordovician period named after Taconic Range in the northern Appalachians, U.S.A.

Tear fault A type of strike slip fault which is vertical but associated with a low-angle overthrust fault towards the hanging wall. The displecement of over thrust side is horizontal (Fig. 144).

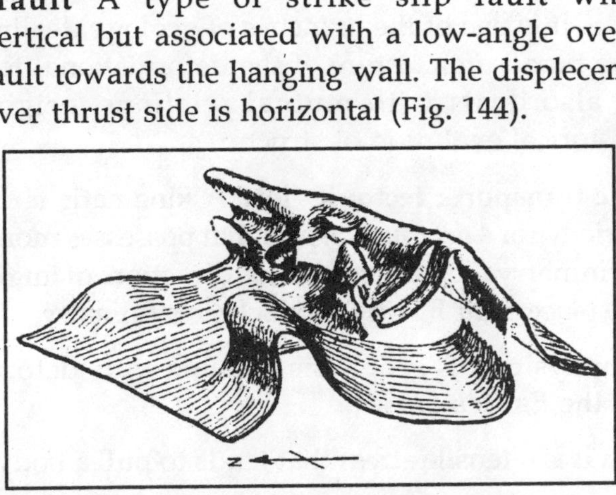

Fig. 144. (After Goguel in Hills, 1973).

Tectofacies Tectonically interpreted lithofacies for "a group of strata of different tectonic aspect from laterally equivalent strata" *Krumbein & Sloss (1951)* included in it the laterally varying tectonic aspects of a stratigraphic unit".

Tectogenesis *(orogeny).*

Tectonic : geotectonic It pertains to earth forces resulting into structural features.

Tectonic breccia It is associated with brittle deformation.

Tectonic conglomerate *(crush conglomerate)*.

Tectonic creep The slow, and continuous movement on a fault.

Tectonic framework The set up of subsiding, stable, and rising tectonic elements on a regional scale.

Tectonic map A map which depicts the structural architecture showing folds, faults and regional dips. It also depicts the history of the varying ages of the structures and rocks.

Tectonics It is one of the branches of geology dealing with the broad architecture of the deformational features, It also deals with mutual relations, origin, and historical evolution of structures.

Tectonic transport : tectonic flow A kinematic term used strictly for a *deformation plan* that possesses monoclinic symmetry. It also reresents direction of maximum displacement having monoclinic symmetry.

Tectonophysics It is geophysics dealing with deformation of the Earth's crust.

Tension It is a tensile stress that tends to pull a body apart.

Tension fault A fault formed by tension.

Tension fracture A fracture which is formed due to tensional stress in a rock. (Fig. 145).

Tension gash A feather like tension fracture alongwhich the walls have been pulled apart. Tension gashes have an én échelon pattern and may occur diagonally in fault zones. (Fig. 146).

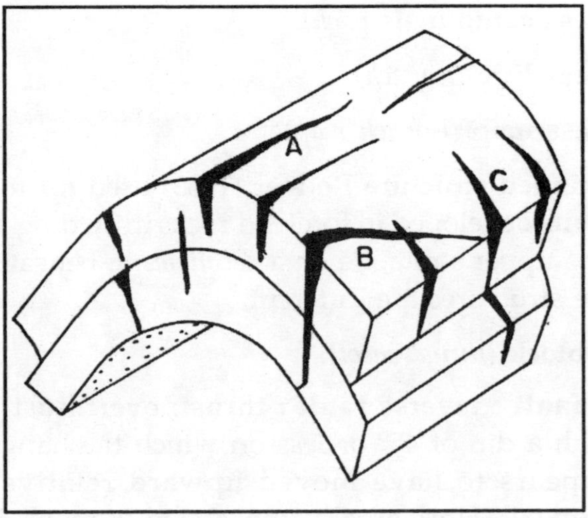

Fig. 145. A-Tension Joint (Fracture) B-Oblique Joint (Fracture) C- Cross Joint (Fracture)

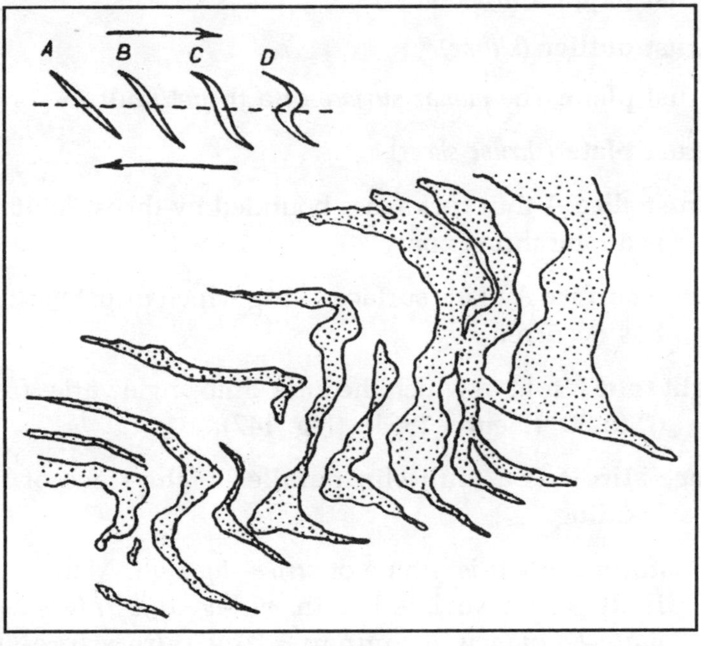

Fig.146. Sigmoidal *(en echelon)* Quartz veins in Nagthat Quartzite showing various stages for their development.

Thickness contour *(isopach).*

Thickness line *(isopach).*

Thickness map *(isopach map).*

Thin-skinned structure Rodger (1963) used for folds and faults developed in foreland rocks in an orogenic belt. The upper strata, lie on a *décollement,* beneath which the structures are different.

Thrust block *(thrust sheet).*

Thrust fault : reverse fault : thrust; overthrust A fault with a dip of 45° or less on which the hangingwall appears to have moved upward relative to the footwall. Horizontal compression is more than the vertical displacement.

Thrust nappe *(thrust sheet).*

Thrust outlier *(klippe).*

Thrust plane The *planar surface* of a thrust fault.

Thrust plate *(thrust sheet).*

Thrust slice A thin rock mass bounded by thrust faults. It is a long thrust zone.

Thrust surface A plane surface, along which thrust faulting has occurred.

Tight fold A fold in which the inter-limb angle varies from 0° to 30° (Fleuty, 1964). (Fig. 147).

Trace-slip It is a fault slip parallel to the trace of the bedding.

Transform fault It is a type of *strike-slip fault.* Many such faults are associated with *mid-oceanic ridges* or a *plate boundary* accompanied by pure stirke-slip displacement (Wilson, 1965). (Fig. 148).

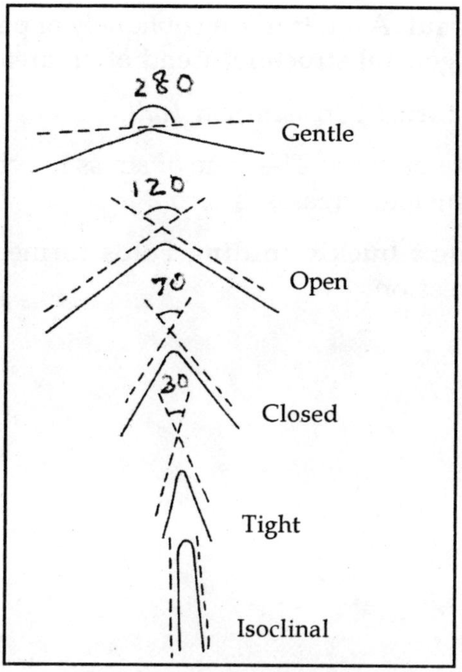

Fig. 147. Tight folds

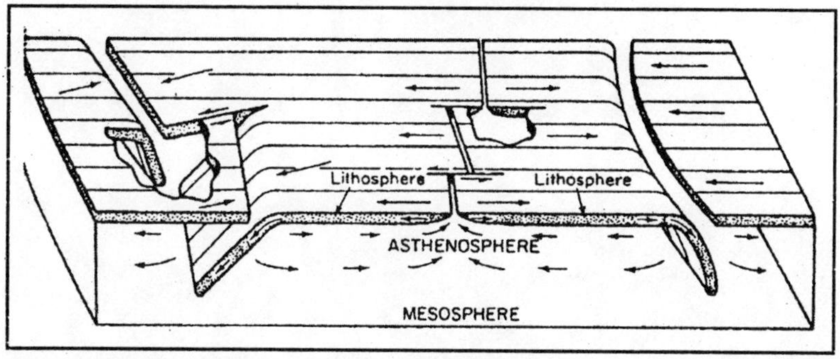

Fig. 148. Plate tectonics and sea-floor spreading. The lithosphere plunges downward along the Benioff zone. (After Billings, 3rd edition, 2000).

Translational fault : translatory fault A fault in which translational movement is present and rotational component is absent (Dennis, 1967).

Transverse fault A fault striking obliquely or perpendicular to the general structural trend of an area.

Transverse thrust : *transcurrent fault.*

Triaxial state of stress The state of stress in which none of the principal stresses is zero.

True folding : buckle folding Folds formed by lateral compression.

□□□

U

Uinta structure An upwarping in the form of a regional, flattened, anticlinal flexure in which the basement-rocks are exposed in the core. It has been named after the Uinta Mountains of Utah.

Ultramylonite An ultra-crushed variety of *mylonite,* which is homogeneous and dense with less of parallel structures (Quensel, 1916).

Unconformity (a) It is break or gap in the geologic sedimentation where a rock unit is overlain by another. It is an interruption in the depositional sequence of sedimentary rocks or a break between eroded igneous rocks and younger sedimentary strata. The structural relationship between rock strata is characterized by absence of continuity in deposition and of nondeposition (subaerial or subaqueous), prior to the deposition of the younger beds. The relationship of the younger overlying stratum does not "conform" to the dip and strike of the older underlying rocks e.g., an angular unconformity. Common types of unconformities are *nonconformity; angular unconformity; disconformity; and paraconformity* (Fig. 149).

Underthrust fault A type of fault in which under thrust rock mass is more active than overthrust part. (Fig. 150).

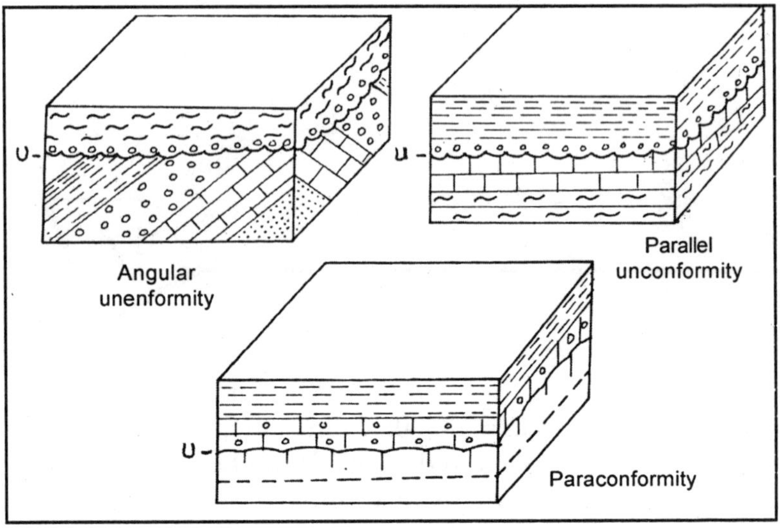

Fig. 149

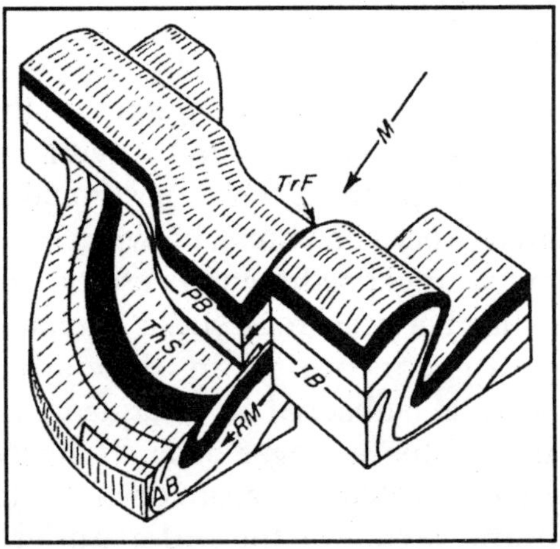

Fig. 150. Underthrust fault passing into overturned fold at one end and broken by a tear fault at the other end. M=direction of movement of underthrust fault block; RM = relative movement of hanging and footwalls of thrust fault along tear fault, shown by length and direction of arrows; TrF = tear fault; ThS = surface of footwall of thrust fault; IB = inactive block of tear fault or region of less intense deformation; PB=passive or hangingwall block of under thrust fault; AB = active or footwall block of underthrust fault. *(From Lovering 1932.)*

Unicline An obsolete term for *monocline.*

Universal stage : U-stage; Federov stage A *stage [optics]* of four, or five axes, attached to the rotating stage of microscope. The thin section of a deformed rock can be studied in all the directions to determine their optical properties.

Uplift [tect] A structurally high area in the crust, produced by uprising movements as is seen in a dome also.

Uplimb thrust fault : flexuralslip thrust fault A contraction fault developed on the limb of an anticline in which the direction of tectonic transport is uplimb and the fault dips initially in the same direction as the limb but at a steeper angle. Continued rotation of the limb towards the vertical side results in associated limb-contraction faults dipping toward the fold axis (*Perry & DeWitt, 1977*).

Upthrow The upthrown side of a fault. charactrized by upward vertical displacement.

U-stage *(universal stage).*

Uttarkashi Earthquake According to Khattri *et al.,* (1989) and Khattri (1992) most of the earthquakes in Uttaranchal originated at a depth of less than 13 km and ascribed them mainly due to underthrusting of the Indian Shield in Lower and Higher Himalayan region (Fig. 151).

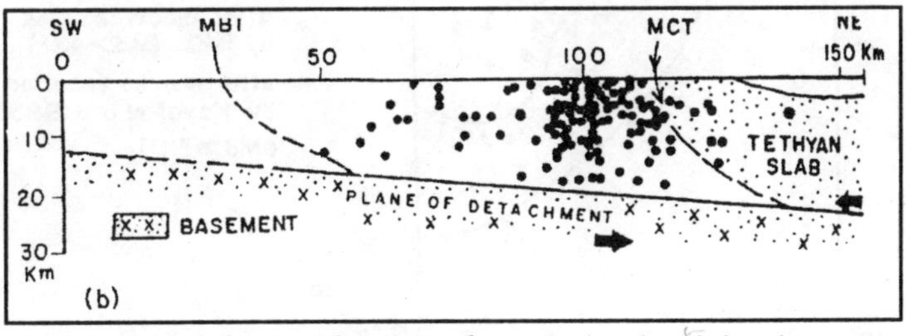

Fig. 151. Model for explaining the microseismicity in terms of underthrusting of the Indian continental crust underneath the Himalaya (Khattri *et al.,* 1989).

The Uttarkashi and earthquakes in recent years in Uttaranchal are of great geological significance.

This earthquake occurred on the 20th October, 1991. Its epicenter was 30.75° N, 78.86° E. Rastogi (2000), Rastogi and Chadha (1995,) Kayal (1996) opined that the focal depth of the earthquake was 12 km with a magnitude of 6.6 on Richter Scale. It took place at 0.2 (H): 53(M). The maximum intensity of the earthquakes was VIII+ on MM scale. It was followed by as many as 40 after shocks. The peak ground accelerations were highest at Bhatwari and Uttarakashi (30% of g). The maximum duration of strong motion was more than 1% g for 40 seconds. The major and minor shocks were confined to the Main Central Thrust Zone indicating about reactivation of the thrust. Kayal *et al.*, (1995) and Kayal (2001) gave cross-sections (Fig. 152) of the aftershocks and ascribed the Uttarakashi earthquake due to rupture of the Indian plate along a detachment surface (underthrusting) underneath the Lesser Himalaya. Seeber and Armbruster (1981) also hold a similar view. The Higher Himalaya underlies the Himalayan

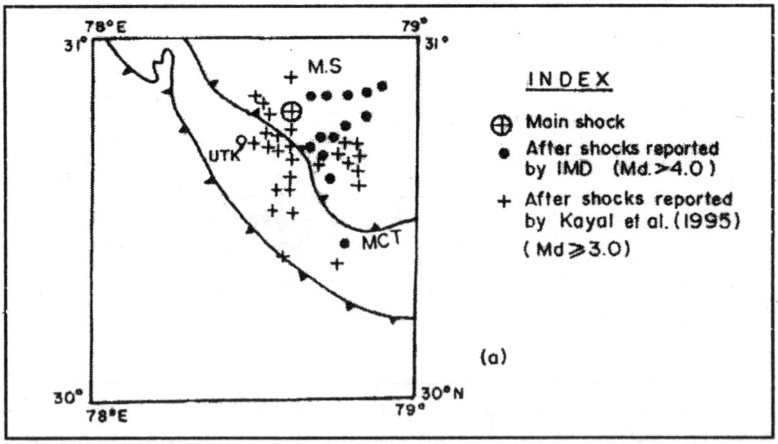

Fig. 152. The figure shows the locations of the Uttarkashi area.

sedimentary wedge and due to rupture of Indian plate associated with northward movement were the main causes for the rise to earthquakes along the Main Boundary and Main Central Thrusts in Uttarakhand.

❏❏❏

V

Variscan orogeny The late Palaeozoic orogeny extending through the Carboniferous and Permian in Europe. It is synonymous with the *Hercynian orogeny*.

Vertical dip slip *(vertical dip)*.

Vertical shift The vertical component of the shift.

Vertical slip : vertical dip slip It is a vertical component of the net slip.

Vertical tectonics The tectonics in which vertical movements of crustal rocks of varying specific gravity control the topographic relief and structures.

Vortex A vertical, cylindrical fold formed in incompetent rock by deformation during deep-zone orogeny (Wynne-Edwards. 1957).

□□□

W

Wall : fault wall The rock unit constituting the hangingwall or footwall.

Wall rock The rock unit comprising the fault-wall.

Warp An upward or downward regional flexure or bend of the crustal rocks.

Warped fault It is a lightly folded thrust fault.

Warping [tect] The phenomena of flexing or bending of the rocks on a regional scale. It can be *upwarping* or *downwarping*.

Wiechert-Gutenberg discontinuity *(Gutenberg discontinuity).*

Window : fenster When younger rocks are exposed and due to erosion are surrounnded by older rocks.

World rift system A major tectonic element of the Earth (e.g., the Mid-Atlantic Ridge). A term used to disribe a group of midoceanic ridges and associated rift-valleys.

Wrench fault : torsion fault A *lateral fault* more or less characterized by a vertical fault surface.

□□□

Y

Young [struc geol] The tem represents the younger aspect of one formation toward another formation. Bailey (1934) used this term.

□□□

Z

Zigzag fold A fold formed by kinking in which limbs of are of unequal length.

Zone of fracture and plastic flow The stronger rocks are deformed by fracturing while the weaker rocks are deformed by plastic flow. These zones occur at intermediate depths of the crust.

□□□

References

Acharyya, S.K., 2005. The Himalaya and the Indian - Asian Plates interactions, In : Saklani P.S. (ed.), Himalaya (Geological Aspects) 2, 275-293, Satish Serial Pub. House, Delhi.

Alvarez, Walter, Engelder, Terry, and Geiser, P.A. 1978. Classification of solution cleavage in pelagic limestones. Geology, v.6, 263-266.

Auden, J.B., 1937. Structure of the Himalaya in Garhwal. Rec. Geol. Surv Ind., 71:407-433; Calcutta.

Bahuguna, V.K. and Saklani, P.S., 1988. Tectonics of Main Central Thrust in Garhwal Himalaya, U.P., - J. Geol. Soc. Ind., 31(2): 197-209; Bangalore.

Bailey, E.B., 1934. West Highlines tectonics, Lochleven to Glen Roy., Q.J.G. S., London, V.30, 462-525.

Bajpai, V.N. and Kandwal, K.N., 2005. Synsedimentary Tectonics in the Western Part of the Marginal Gangetic Alluvial Terrain and its implication for Aquifer Disposition: Case study from Delhi-Faridabad-Bulandshahr, India, Special publication of the Palaeontological society of India, No. 2, 245-260.

Bhatt, S.C. and Saklani, P.S., 1990. Kinematic framework of heterogeneous deformation within Pratapnagar thrust sheet, Bhagirathi Valley, in Lesser Garhwal Himalaya, U.P., Jour. Geol. Soc. India, 36 (3), 247-261.

Billings, M.P. 2000. Structural Geology. N.J.: Prentice-Hall. 606p.

Boyer, S.E. and Elliott, D., 1982. Thrust Systems. - Bull. Am. Assoc. Pet. Geol. 66(9): 1196-1230; Tulsa.

Bucher, W.H., 1955. Deformation in orogenic belts., Geol. Soc. Am. Mem. 49, 113 p.

Burg, J.P., 2006. Two orogenic systems in the Himalaya : evidence and consequences., I-33. In : Saklani, P.S. (Ed.), Himalaya (Geological Aspects), 4, Satish Serial Pub. House, Delhi.

Butler, R.W.H., 1982. The terminology of structures in thrust belts. Jour. Struct. Geol., 4(3):239-245.

Carey, S.W., 1958. A tectonic approach to contineental drift. In: Carey, S.W., convener. Continental drift symposium. Hobart: University of Tasmania, Gelogy Dept. 177-355.

Catlos, E.J., Dubey, C.S., Harrison, T.M. and Edwards, M.A., 2004. Late Miocene Movement within the Himalayan Main Central Thrust Shear Zone, Sikkim,Northeast India, Jour. Met. Geol., 22, 207-226.

Cloos, E., 1946. Lineation : A critical review and annotated bibliography, Geol. Soc. Am. Mem. 18, 122 p.

Cooper, M.A., 1981. The internal geometry of nappes-criteria for models of emplacement. In: 'Thrust and Nappe Tectonics' (Edited by McClay, K.R. and Prince, N.J.). Special Publ. Geol. Soc. London, 9:335-344.

Dahlstrom, C.D.A., 1969. Balanced cross-sections. - Can. J., Earth Sci., 6:743-746; Ottawa.

Dahlstrom, C.D.A., 1970. Structural geology in the eastern margin of the Canadian Rock Mountains. Bull. Can. Petrol. Geol., 18:332-406.

Dana, J.D., 1873, On some results of the Earth's contraction from cooling including a discussion of the origin of mountains and the nature of the earth's interior, Am. Jour. Sci. V. 5, 423-443.

Dennis, J.G., ed., 1967. International Tectonic Dictionary, A.A.P.G., Mem., 7, 196 p.

Dennis, J.G., and Atwater, T.M., 1974. Terminology of geodynamics. American Association of Petroleum Geologist. Bull. 58, 1030-1036.

Dewey, J.F., 1969. Continental Margins- a model for conversion of Atlantic type to Andean type, Earth Planet, Sci. Lett. V. 6, 189-197.

Dezes, P., 1999, Tectonic and metamonrphic evolution of the Central Himalayan domain in southeast Zanskar (Kashmir, India), Mem. Geol., 32, Lausanne.

Douglas, R.J.W., 1950. Callum Creek, Langford Creek, and Gap map areas, Alberta. Geological Survey of Canada. Memoir 255. 124p.

Dubey, A.K. and Jayangondaperumal, 2005. Satengal klippe in the Mussoorie syncline, Lesser Himalaya, Evidence from field and Model deformation studies, In : Saklani, P.S., 2005, Himalaya (Geological Aspects) 3, 203-222, Satish Serial Pub. House, Delhi.

Emmons, W.H., 1933. On the mechanism of the deposition of certain metalliferous lode systems associated with granitic batholiths. In: The Committee on the Lindgren Volume, ed. Ore deposits of the western states. New York: American Institute of Mining and Metallurgical Engineers. 797p.

Epard, J.L., Steck, A., Vannay, J.C. and Hunziker, J., 1995, Tertiary Himalayan Structures and Metamorphism in Kulu Valley (Mandi-Khoksar transect of the Western Himalaya), Shikar Beh Nappe and Crystalline nappe, Schw. Min. Pet. Mit., 75, 59-84.

Fleuty, M.J., 1964. The description of folds. Geologists' Association. Proceedings, v.75, p.461-492.

Friedman, M., and others, 1976. Experimental folding of rocks under confining pressure: Part-III. Faulted drape folds in multilithology layered specimens. Geological Society of America. Bulletin, v.87, p.1049-1066.

Gansser, A., 1964. The Geology of the Himalayas, Wiley Interscience, 289 p.

Gansser, A., 1993. Facts and theories of the Himalaya. Jour. Geol. Soc. Ind., 41, 487-508.

Gilbert, G.K., 1890. Lake Bonneville. U.S. Geological Survey. Monograph 1. 438p.

Girard, M., 2001, Metamorphism and tectonics of the transition between nonmetamorphic tethyan Himalayan sediments and North Himalayan Crystalline Zone & Rupshu area, Ladakh, NW India, Mem. Geol. (Lausanne), 35.

Glassner, M.F. and Teichert, C, 1947. Geosynclines: a fundamental concept of Geology, Am. Jour. Sci., V. 245, 465-42, 571-591.

Gussow, W. C., 1962. Energy source of intrusive masses. Royal Society of Canada. Trans., v. 56, series 3, sec. 3, p. 1-19.

Haarmann, E., 1930. Die Oszillationstheorie; eine erklarung der krustenbewegungen von erder und mond. Stuttgart: Enke, 260p.

Hall J., 1843. Geology of New York Part IV, Survey of the fourth geological district, Albony, Carroll and Cook, 683 p.

Hills, E.S., 1973. Elements of Structural Geology, 483 p. Methew & Co., Ltd., London.

Holmes, A., 1965. Principles of physical geology, 1288 p. Eng. Lang. Book Soc. & Nelson London.

Hunt, C.B., and others, 1953. Geology and geography of the Henry Mountains region, Utah. U.S. Geological Survey. Professional Paper 228. 234p.

Jones, P.B., 1971. Folded faults and sequence of thrusting in Alberta foothills. American Association of Petroleum Geologists. Bulletin, v.55, p.293-306.

Kay, G.M., 1947. Geosynclinal nomenclature and the craton. American Association of Petroleum Geologists. Bulletin, v.31, no.7, p.1287-1293.

Kayal, J.R., 1996, Earthquake source parameters in north east India: A review, Jour. Him. Geol., 17, 53-60.

Kayal, J.R., 2001, Micro earthquake activity in some parts of the Himalaya and the tectonic model, Tectonophysics, 339, 331-351.

Kayal, J.R., De, R., Chakraborty, R. and Karunakar, G., 2003. Aftershock Sequence of Chamoli Earthquake of March 29, 1999 in Garhwal Himalaya, Bull. Seis. Soc. Am., 93, 109-113.

Kerr, P.F., and Kopp, O.C., 1958. Salt-dome breccia. American Association of Petroleum Geologists. Bulletin, v.42, p.548-560.

Khattri, K.N., 1992. Local Seismic investigations in Garhwal Himalaya, Geol. Soc. Ind. Mem., 23, 45-66.

Khattri, K.N., Chander, R., Gaur, V.K., Sarkar, I. and Kumar, S., 1989. New Seismological results on the tectonics of Garhwal Himalaya, Earth. Planet. Sci. Lett., 98, 91-100.

Krumbein, W.C., and Sloss, L.L., 1951. Stratigraphy and sedimentation. San Francisco: Freeman. 497p.

Krummenacher et al., 1978. Petrology, metamorphism and K-AR age determinations in eastern nepal, Tectonic Geology of the Himalaya In : Saklani, P.S. (ed.) 151-158, Today and Tomorrows, Delhi.

Lange et al., 1966. General Geology, For. Lan. Pub. House Moscow, 299 p.

Lahee, F.H., 1971. Field Geology, 6th ed., McGraw Hill, Tokyo, 926p.

Leith, C.K., 1905. Rock cleavage. U.S. Geological Survey. Bulletin 239. 216p.

Lisle, R.J., 1988. Geological Structures and Maps : A practical guide book, Pergamon, Oxford, 150 p.

Longwell, C.R., 1951. Megabraccia developed down slope from large faults, Am. Jour. Sci., v. 249, 343-355.

Lovering, T.S., 1932. Field evidence to distinguish overthrusting from underthrusting. Jour. Geol, 40, 651-663.

Malvern, L.E., 1969. Introduction to the mechanics of a continuous medium. Englewood Cliffs, N.J.: Prentice-Hall. 713p.

McBride, E.F. and Yeakel, L.S., 1963. Relationship between parting lineation and rock fabric. Jour. Sed. Petr., v. 33, 779-782.

McKee, E.D., and Weir, G.W., 1953. Terminology for stratification and cross-stratification in sedimentary rocks. Geological Society of America. Bulletin, v.64, p.381-389.

Naumann, C.F., 1858. Lehrbuch der Geognosie. Bd. 1. Leipzig: Wilhelm Engelmann. 960p.

Nettleton L.L., 1934. Fluid mechanics of salt domes A.A.P.G. Bull;, 18, 1175-1204.

Norris, D.K., 1958. Structural conditions in Canadian coal mines. Geological Survey of Canada. Bulletin 44. 54p.

Norris, D.K., 1960. Microtectonics of the Kootenay Formation near Fernite, British Columbia. Bulletin of Canadian Petroleum Geology Survey, v.12, p.383-398.

Norton, W.H., 1971. A classification of breccias. Journal of Geology, v.25, p.160-194.

Patriat, P. and Achache, J., 1984. India-Eurasia Collision chronology has implications for crustal shortening and driving mechanism of plates. Nature, 311, 615-621.

Perry, W. J. and Dewitt, W., 1977. A field guide to thin skinned tectonics the central Appliachians A.A.P.G., 54 p.

Peyve, A.V., and Sinitzyn, V.M., 1950. Certain problems fondamentaux de la doctrine des géosynclinaux. Akademiya Nauk USSSR. Izvestiya, Seriya Geologicheskaya, v.4, p.28-52.

Quensel, P., 1916. Zur Kenntnis der Mylonitbildung, erläutert an Material aus dem Kebnkaisegebite. Uppsala. Universitet. Geologiska Institut. Bulletin, v.15, p.91-116.

Ramberg, H., 1955. Nature and Experimental boudinage and pinch and swell structures, Jour. Geol. v. 63, 512-526.

Ramsay, J.G., 1967. Folding and Fracturing of rocks. McGraw Hills, New York, 568 p.

Rastogi, B.K. and Chadha, R.K., 1995. Intensity and isoseismals of Uttarkashi earthquake of Oct. 20, 1991, Geol. Soc. Ind. Mem., 30, 19-24.

Rastogi, B.K., 2000. Chamoli-earthquake of magnitude 6-6. March, 29, 1999, Jour. Geol. Soc. Ind., 55, 505-514.

Riedel, W., 1929. Zur Mechanik geologischer Brucherscheinungen. Zen. Min. Geol. Palaeon., 354-368.

Robyr, M., Vannay, J.C., Epward, J.H. and State, A., 2002. Thrusting, extension and doming during polyphase tectonometamorphic evolution of the High Himalaya Crystalline Zone, NW India, Jour. Asian Earth Sci., 21, 221-239.

Rodger, John, 1963. Mechanics of Appalachian foreland folding in Pennsylvania and West Virginia. America Association of Petroleum Geololgists. Bulletin, v.47, p.1527-1536.

Saklani, P.S., 1993. Geology of Lower Himalaya (Garhwal), 240 p. International Books, Delhi.

Sander, Bruno, 1930. Gefügekunde der Gesteine mit besonderer Berücksichtigung der Tektonite. Vienna: Julius Springer. 352p.

Schardt, H., 1893. Surl'origine des alpes du chablais et du stockhorn en savoie et en suisse, com. reb. sci. acad. 117(21), 707-709.

Schuchert, C., 1923. Sites and natures of the North American. Geosyncline, Soc. Am. Bull., v. 34, 151-260.

Schulp, M., Carter, A., Cosca, M. and Steck, A., 2003, Exhumation history of eartern Ladakh revealed by 40 Ar / 39 Ar and fision track ages, the Indus river, Tso-Morari transect, NW Himalaya, Jour. Geol. Soc. Lond., 160, 1-15.

Schwan, W. 1980. Shortening structures in Eastern and North-western Himalayan rocks. in : Saklani P.S. (Ed.); 80 p. Today & Tomorrow's, Delhi.

Schwan, W. and Saklani, P.S., 1991. Mesoscopic structures along the Bhatwari - Narendranagar traverse Garhwal Uttar Pradesh and their chronologic and kinematic integration of deformation plan of the Himalaya Z. dt. Geol. Ges., 142, 97-113.

Seeber, L. and Armbruster, J.R., 1981, Great detachment earthquakes along the Himalayan arc and longterm forecasting, In : Earthquake prediction, an International Review, Am. Geoph. Union Pub., Ser., 4, 259-275.

Shekhar et al., 2006. Geology and Structure of Srinagar Garhwal. In: Saklani P.S. (ed.), Himalaya (Geological Aspects) Vol. 4, 153-169, Satish Serial Pub., Delhi.

Steck, A., Spring, L., Vannay, J.C., Mason, H., Bucher, H., Stuiz, E., Merchant, R. and Tieche, J.C., 1993, Geological transect across the north western Himalaya, Eastern Ladakh and Asia, Eclo. Geol. Helv., 86, 219-263.

Stille, H.W., 1940. Einführung in den Bau Amerikas. Berlin: Gebrüder Borntraege. 717p.

Stille, Hans, 1930. Über Einsetigkeiten in der germanotypen Tektonik Nordspaniens and Deuschlands. Gesellschaft der Wissenschaften, Geöttingen. Mathematisch-Physikalische Klasse. Nach-richten, 1930, H.3, p.379-397.

Stille, Hans, 1935. Der derzeitige tektonische Erdzustand. Preussische Academie der Wissenschaften, Physikalisch-mathematische Klasse. Sitzungsberichte, v.13, p.179-219.

Stille, Hans, 1936. Present tectonic state of the Earth. American Association of Petroleum Geologists. Bulletin, v.20, p.849-880.

Stokes, W.L., and Varnes, D.J., 1955. Glossary of selected geologic terms, with special reference to their use in engineering. Denver: Colorado Scientific Society. 165p.

Turner, F.J., and Weiss, L.E., 1963. Structuctral analysis of metamorphic tectonites. McGraw-Hill. New York 575p.

Tyrrell, G.W., 1926. The principles of petrography; an introduction to the science of rocks. London : Methuen. 349p.

Umbgrove, J.H.F., 1933. Verschillende Typen von Tertiaire Geosynclinalen in den Indischen Archipel., Leid. Geol. Med., v.5 (1), 33-43.

Valdiya, K.S., 1971. Samanya Bhuvigyan: Kuchh Jwalant Samasyain Hindi Gra. Aca., 267 p.

Valdiya, K.S., 1984. *Aspects of Tectonics,* Tata Mc Graw Hill Pub., 320 p.

Wegener, A., 1912. The origin of the continents and oceans : English translation in 1924, New York, Dutton.

Wentworth, C.K., 1935. The terminology of coarse sediments with notes by P.G.H. Boswell. National Research Council. Division of Geology and Geography. Committee on Sedimentation, Report for 1932-1934, p.225-246. (National Research Council. Bulletin, 98).

Whitten, D.G.A. and Brooks, J.R.V., 1972. The Penguin Dictionary of Geology, 495 p., Penguin Books, London.

Wilson J.G., 1953. Mullion and Rodding structures with Moine series in Scotland, Prof. Geol. Assoc. London, v. 64, 118-151.

Wilson, G., 1961. The tectonic significence of small scale structures and their importance to the geologist in the field. Soc. Geol. Belg., 84, 423-548.

Wilson, J.T., 1968. Static or mobile earth In : Gondwanaland revisited, New evidence for continental drift, Am. Phil. Soc. Proceed, v. 112, 309-320.

Wlison, J.T., 1965. A new class of faults and their bearing on continental drift. Nature, v.207, p.343-347.

Wynne-Edwards, H.R., 1957. Structure of the Westport concordant pluton in the Grenville, Ontario. Journal of Geology, v.65, p.639-649.

□□□

Author Index

T

U

V

W

❑❑❑

Subject Index

B

C

D

E

F

G

H

N

O

P

Q

R

S

❑❑❑